Rainwater Harvesting

Anil Lalwani

Rainwater Harvesting

In Urban Centers within the Hard Rock
Terrain of the Deccan Basalts of India

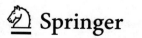 Springer

Anil Lalwani
Well & Water Works
Pune, India

ISBN 978-3-031-11568-4   ISBN 978-3-031-05710-6 (eBook)
https://doi.org/10.1007/978-3-031-05710-6

This Springer imprint is published by the registered company Springer Nature Switzerland AG
The registered company address is: Gewerbestrasse 11, 6330 Cham, Switzerland

This book has been dedicated to my late wife Anuradha who was there with me right till her end, standing behind me and supporting me in all my endeavors and that really has helped in making me what I am today, and also to all my colleagues, friends, and clients, who put their trust in my ability in coming up an amicable solution to their water-related issues.

Preface

This book discusses in brief the classifications of rainwater harvesting and the various rainwater harvesting systems that are currently in vogue especially in urban areas. The most important aspect for achieving any groundwater recharge is the availability of sustainable source water; this can be evaluated by analyzing the monsoon rainfall pattern, its frequency, number of rainy days, and maximum rainfall in a day and its variation in space and time.

Rainwater harvesting within the Deccan Basaltic terrain is not really as simple as it is assumed to be. This is not surprising, as the Deccan basalts are one of the most enigmatic rocks and pose a very difficult task where it comes to groundwater exploration and naturally also for groundwater recharge, especially in the urban areas where due to constraint of space one needs to resort to borewells for recharge purpose.

In light of the above-mentioned characteristics of the Deccan basaltic aquifer, it becomes very important to understand the applicability and sustainability aspects of rainwater harvesting as it is being practiced.

This book tries to differentiate the Basaltic aquifers from the other hard rock aquifers and highlights the difficulties in trying to implement rainwater harvesting by groundwater recharge with the basaltic terrain. It also deals with the issue of long-term sustainability of rooftop rainwater to meet the growing demands of fresh water.

It highlights some of the shortcomings in the methodologies used and the requirements for being considered in the various categories of green building rating agencies.

It also highlights the possibility and limitations of dependence of rooftop harvesting in addressing the question of water shortages, which is of common occurrence within most urban centers of India. Further, more clarifications regarding some of the major misconceptions that are currently prevailing regarding rooftop rainwater harvesting especially within the low-capacity aquifers within the hard rock terrain of the Deccan basalts have also been made.

Pune, India   Anil Lalwani

Acknowledgments

I would like to express my gratitude to Well & Water works for providing me an opportunity to work on such interesting and practically oriented endeavor, and my special thanks to Darshana and Bharat Mane.

I would like to thank Shri. Rahul Sakore, of Pushkaraj Consultancy, and Shri. Sujit Dengle, former employee of Marvel Developers, for putting their faith in me and my abilities to come up with suitable solutions in this field. I would like to express my special gratitude to Shri. Suhas Joshee and Rahul Sakore for taking time to prereview this document in its nascent stages and make some good suggestion for improving it.

I would like to take this opportunity to acknowledge Larry W. Mays for the photos. Photo 1.1: Minoan Water Systems on Crete and Photo 1.2: Courtyard used for rainfall harvesting with cisterns (round structures) shown in background to the right at Phaistos, Greece, Diego Delso for Photo 1.3, of Acueductos subterráneos de Cantalloc, Nazca, Perú, Rama's Arrow for Photo 1.4 of the Brick Lined Water Reservoir at Dholavira, kangarotr, for Photo 1.5: A Small Johad in Laporiya village of Rajasthan, India, WikiPanti for Photo 1.6 of Talabs for Irrigation for Plants and Drinking, Purohit for Photo 1.7: Baoris—Drinking water wells, Chand Baori, Gagnon for Photo 1.8: Jhalaras, Tanks Constructed for Religious Purpose.

I would also like to acknowledge IGBC for Table 3.1—Mandatory criteria to arrive at "One-day Rainfall"; Table 3.2—Criteria to arrive at "One-day Rainfall" for additional points; Table 3.3—Criteria to arrive at "One-day Rainfall" for exemplary performance and Table 3.4—In projects with groundwater levels less than 8 m, Criteria to arrive at "One-day Rainfall".

This acknowledgement would be incomplete without the mention of all our clientele and the valuable assistance and discussions I had with them while working on projects.

I wish to thank also my friends, the colleagues, and associates for their advice and valuable help during the compilation of this book; a special mention is needed for my friend Sirje Kreisman who has been encouraging me to sharing my knowledge by way of writing this book.

Last but not least, I owe everything to my Daughter Kanika who was always there to support me and inspire me in times of need.

Pune, India   Anil Lalwani

Contents

Abbreviations

A	Area of the catchment, in square meters
ACWADAM	Advanced Centre for Water Resources Development and Management
ATNESA	Animal Traction Network for Eastern and Southern Africa
BASF	Badische Anilin und Soda Fabrik
BIS	Bureau of Indian Standards
BMC	Billion cubic meters
CCT	Continuous Contour Trenches
CGWA	Central Groundwater Authority
CGWB	Central Groundwater Board, Ministry of Water Resources, India
Col	Colonel
CPWD	Central Public Works Department
DTH	Down-the-hole hammer
FAO	Food and Agriculture Organization
GRIHA	Green Rating for Integrated Habitat Assessment
GSDA	Groundwater Survey and Development Agency
IGBC	Indian Green Building Council
IUZ	Inter Unit Zone
KSCST	Karnataka State Council of Science and Technology
m^3	Cubic meter
mm	Millimeter
MUD	Ministry of Urban Development, India
NBC	National Building Code
NDTV Ltd.	New Delhi Television Limited
NGO	Non-Governmental Organization
NITI Aayog	National Institute for Transforming India
NIUA	National Institute of Urban Affairs
R&D	Research and Development
retd	Retired
Rf	Amount of rainfall
RHP	Rainwater Harvesting Potential

SCT	Staggered Contour Trenches
SGS	Shallow Groundwater System
SIWI	Stockholm International Water Institute
Sq. m.	Square meter
STP	Sewage Treatment Plants
UNEP	United Nations Environment Program
WSIDDR	Water and Sanitation in International Development and Disaster Relief

Chapter 1
Introduction

Abstract Availability of good quality water is one of the essential factors for any kind of sustainable development. After reaching the road block in terms of lack of space in metros to spread laterally seems to have prompted the administration to reach toward the sky to meet the ever-increasing demand for living space. This upward trend in construction has given rise to high-rise multi-storied building being built in all urban metros across the world. This in turn has led to an increase in the population density which is concentrated in and around such metro cities, and with this increase in population density comes the extra demand of fresh water.

Keywords Sustainable development · Fresh water demand · Urbanization · Rainwater harvesting · Cultural traditions

In India, a growing population and the migration of large numbers of people from rural areas to cities in the last few decades have led to an ever-increasing demand for fresh water in the urban metros. Most cities are unable to meet this demand with the existing infrastructure that has been there since decades which unfortunately has not been upgraded to meet this growing demand and hence are trying to come up with solutions to augment the supply of fresh water to its citizens.

A lack of free and cheap space in urban areas makes it impossible to construct common large surface reservoirs, which can store water in an appreciable quantity to mitigate the water problems faced by such large populations. It is on this background that the traditional practices of rainwater harvesting are being revived; with the government and private organizations all vouching for urban rainwater harvesting as the solution for meeting this ever-increasing demand for water. Urban rooftop rainwater harvesting is being seen as a possible solution to help the population to get sufficient water for their needs—particularly during the hot dry summer months when shortages of water are usually most severe.

Rooftop rainwater harvesting, well it seems that everyone in the twenty-first century seems to be talking about it. The city planners, the government agencies, the engineers, the architects, the corporates, and the NGOs everyone seems to know what it is and how it is to be implemented. Moreover, everyone is also being made to think that it is the only solution in overcoming the water shortages being faced by a large majority of people living in the fringes or the neo-development areas of urban metros of India and in fast-growing urban metros of India.

Rainwater harvesting is therefore being considered as a social responsibility of every good citizen, with a single goal that being to ensure the health—in terms of both the quality and quantity—of the natural groundwater system. Most importantly, this will mean that the groundwater extracted during the periods of little or no rainfall will have to be replenished by using rainwater when it is raining, which is what normally happens within the natural hydrogeological cycle.

Due to urbanization, there is an obvious increase in surface runoff which is caused due to the reduction of open natural ground which is being converted to treated hard play grounds or concrete surfaces and roads which in turn obstructs the water from infiltrating into the ground. This change in land use can lead to an increase in surface runoff which can be as high as 40% or more, and thereby resulting in a decrease in natural groundwater recharge at the same time, the resultant increased surface runoff causes an increase in the load on the existing storm water system which had been laid done decades ago. Ideally speaking, rainwater harvesting should be aimed at meeting this shortfall of reduced recharge, anything else would only result in disturbing the natural ecological balance, and anything more would probably be rejected by nature itself.

The government and the NGOs are all are vouching for urban rainwater harvesting as the solution for meeting the ever-increasing demand of water due to growing urban centers, which is noticeable especially during the summer months. It is during this period that the already overexploited groundwater system which supplements/compliments the municipal supply is inadequate to meet the daily demands. In spite of all the publicity, government initiatives by way of passing legislation of making rainwater harvesting compulsory and offering discounts in property taxes, etc. have really done very little to overcome the water woes of the masses at large.

Well, with all these efforts being made by the government agencies to improve the water situation in urban centers during the summer months has really not been much of help, and hence, it makes one wonder, as to what exactly is rainwater harvesting? As per the definition in the Wikipedia (2021), *"Rainwater harvesting" is a type of harvest in which the rainwater is collected from roof like surfaces and stored for the future use, rather than allowing them to runoff. Rainwater can be collected from rivers or roofs and redirected to a deep pit (well, shaft, or borehole), aquifer, a reservoir with percolation, or collected from dew or fog with nets or other tools and stored for use in the future.*

Rainwater harvesting is nothing new and had been practiced in ancient times too, its revival in the present times is due to the water shortages that are being faced by vast majority of people living in urban centers, and hence, rainwater harvesting is again gathering a lot of significance as a modern, water-saving, and simple technology. Especially in the light of the NITI Aayog Kant (2018), "The Composite Water Management Index" report by Kant which states that—*India is facing the vilest water crisis in its history. The report predicts, 21 Indian cities, including major metros like Delhi, Chennai, Hyderabad, and Bangalore will run out of groundwater by 2020*

Photo 1.1 Minoan water systems on Crete. Stepped water channel and sedimentation (desilting basin). Along the stairway is a small channel (for rainwater collection) that conveys rainwater down steam to the sedimentation tank or basin at **Knossos**, Greece (Photo: Mays 2012)

rainwater which is the only source of pure water rainwater harvesting seems to be the only forward.

Drinking water being supplemented through rainwater harvesting in urban areas, especially in the semi-arid areas, dates back in history right from the late Neolithic to early Bronze Age period. People in Mesopotamia (e.g., today Iraq and Jordan) realized that life is not possible without water and hence resorted to harvesting rainwater to meet their supply of water. Such water supply systems are known from Minoan Crete and in the Indus Valley ca. 3rd millennium BC (Angelakis 2016) (Photos 1.1 and 1.2).

Cowie (2018), states that the use of cisterns to store rainwater can be traced back to the Neolithic Age. He further goes on to state that the emerging water management techniques used in farming yielded waterproof lime plaster cisterns built in the floors of houses in villages of "the Levant" (a large area in Southwest Asia, south of the Taurus Mountains, between the Mediterranean Sea to the west and Mesopotamia in the east), dating back to 4000 BC. The desertified Peruvian Valleys were made livable as early as 1500 years ago due to Advanced Hydraulic Engineering Miller (2016) (Photo 1.3).

Even though rainwater harvesting was greatly developed around 2000 BC in India, China, and Mesopotamia, it was formalized in ancient Rome (Cowie 2018).

Photo 1.2 Courtyard used for rainfall harvesting with cisterns (round structures) shown in background to the right at **Phaistos, Greece** (Photo: Mays 2012)

Photo 1.3 Cantalloc subterranean aqueducts, Nazca, Peru (Photo: Delso 2015)

Photo 1.4 Brick-lined water reservoir at Dholavira (Rann of Kutch)—Harappan Civilization dated 2500–1900 BC (Photo: Rama's Arrow 2006)

According to Gupta and Agrawal (2015), India is described as a country which has very deep historical roots and strong cultural traditions. These, according to them, are reflected in its social fabric and institutions of community life. Some of these traditions, according to them, have evolved and developed thousands of years ago and have played an important role in different spheres of life. One such important tradition among these is the tradition of collecting, storing, and preserving water for various uses.

As the population increased, settlements developed into towns and cities and agriculture expanded, techniques were developed to augment water availability by collecting and storing rainwater, tapping hill and underground springs and water from snow and glacier melt, etc. to meet the growing demand for water.

The first major human settlements in India was the Indus Valley (3000–1500 BC) which came up in the in the North and Western India. It had water systems in them; the evidence of its existence can be found in the literature of this period. Archeological findings indicate existence of irrigation and drinking water supply systems from a large number of brick-lined wells. In Dholavira, which is an important site of Indus Valley, there is evidence of several reservoirs for collection of rainwater (Photo 1.4).

Similar evidences are found at Mohanjodaro and Harappa. In Lothal (Gujarat), Inamgaon (Maharashtra), and other places in North and Western India, small bunds were built by the local people to store rainwater for irrigation and drinking. The Arthashastra of Kautilya gives an extensive account of dams and bunds that were built for irrigation during the period of the Mauryan Empire (321–185 BC) (Gupta and Agarwal 2015).

Photo 1.5 Small Johad in Laporiya village of Rajasthan, India (Photo: Singh 2018)

Pal (2016), on "thebetterindia" website describes the ancient Indian water capturing systems which archeologists have identified, viz., Johads, Talabs, Baoris, Jhalaras, etc.

Johads: Dams that captured rainwater (Photo 1.5).

Talabs: Reservoirs that provided irrigation for plants and drinking (Photo 1.6).

Baoris: Wells in the ground for drinking water (Photo 1.7).

Jhalaras: Specially constructed tanks used for religious purpose (Photo 1.8).

According to Heggen (2000), in the last few decades, there has been an increasing interest in the use of harvested water, with an estimated 100,000,000 people worldwide currently utilizing a rainwater system of some description.

In their concluding remarks, Yannopoulos et al. (2017) point to the fact that, in many countries of Europe, Asia, Africa, America, and Australia, rainwater harvesting is obligatory not only to address water scarcity, but to hold storm water and to reduce flood risks. They further go on to say without doubt, "our past can teach us a lot". They are also of the opinion that historical studies on rainwater harvesting, collection, and storage technologies help us in understanding the possible responses of modern societies to the future sustainable management of water resources. Naturally, as rain being the only source of fresh water, harvesting rainwater gains its rightful place of importance for ensuring adequate water for the future growth of human habitation.

Photo 1.6 Talabs for irrigation for plants and drinking—The Gadisar, a manmade lake at Jaisalmer, constructed in 1367 (Photo: WikiPanti 2011)

Photo 1.7 Baoris—drinking water wells, Chand Baori, the deepest step well in the world, history, Abhaneri Village, Rajasthan (Photo: Purohit 2019)

Photo 1.8 Jhalaras, tanks constructed for religious purpose, The Rani ki vav, Patan, Gujarat (Photo: Gagnon 2013)

References

Angelakis AN (2016) Evolution of rainwater harvesting and use in Crete, Hellas through the millennia. Water Sci Technol: Water Supply 16(6):1624–1638

Cowie A (2018) Ancient rainwater harvesting: it fell from the sky and became worshiped by every civilization. Ancient Origins. https://www.ancient-origins.net/history-ancient-traditions/ancient-rainwater-harvesting-0010904. Accessed 24 October 2020

Delso D (2015) Acueductos subterráneos de Cantalloc, Nazca, Perú, DD 02.JPG, Wikimedia commons contributors, 'File: Acueductos subterráneos de Cantalloc, Nazca, Perú, 29 July, Wikimedia Commons, the free media repository. https://commons.wikimedia.org/wiki/File:Acueductos_subterráneos_de_Cantalloc,_Nazca,_Perú,_2015-07-29,_DD_02.JPG. Accessed 24 June 2021

Gagnon B (2013) Wikimedia commons contributors, 'File: Rani ki vav 02.jpg', Wikimedia commons, the free media repository, 22 September 2020, 04:32 UTC. https://commons.wikimedia.org/w/index.php?title=File:Rani_ki_vav_02.jpg&oldid=467500156. Accessed 11 June 2021

Gupta RK, Agrawal RK (2015) Rainwater harvesting in ancient times and its sustainable modern techniques. ICID2015, pp 1–5. https://icid2015.sciencesconf.org/74834/Paper_for_Submission_to_ICID.pdf

Heggen RJ (2000) Rainwater catchment and the challenges of sustainable development. Water Sci Technol 42(1–2):141–145

Kant A (2018) Composite water resources management—performance of states. National Institute for Transforming India (NITI Aayog), Government of India, New Delhi, Report, 179

Mays WL (2012) Minoan Water Systems on Crete https://ancientwatertechnologies.com/2012/12/26/minoan-water-system-at-tylissos-crete/comment-page-1/. Accessed 21 October 2020

Miller M (2016) Advanced hydraulic engineering made desertified Peruvian Valleys Livable 1,500 years ago. Article at https://www.ancient-origins.net/ancient-places-americas/advanced-hydraulic-engineering-made-desertified-peruvian-valleys-livable-020979. Accessed 15th July 2021

Purohit T (2019) Wikimedia commons contributors, 'File: Ancient way of water presservation.jpg', Wikimedia commons, the free media repository. https://commons.wikimedia.org/w/index.php?title=File:Ancient_way_of_water_presservation.jpg&oldid=493530436. Accessed 18 October 2020

Rama's Arrow (2006) Wikimedia commons contributors, 'File: Dholavira1.JPG', Wikimedia commons, the free media repository, 27 October 2020, 06:38 UTC. https://commons.wikimedia.org/w/index.php?title=File:Dholavira1.JPG&oldid=503481435. Accessed 27 October 2020

Sanchari P (2016) Modern India can learn a lot from these 20 traditional water conservation systems. https://www.thebetterindia.com/61757/traditional-water-conservation-systems-india/. Accessed 15 July 2021

Singh KA (2018) Wikimedia commons contributors, 'File: A Nadi (small johad) in village Laporiya, Rajasthan.jpg', Wikimedia commons, the free media repository, 2 May 2021, 03:01 UTC. https://commons.wikimedia.org/w/index.php?title=File:A_Nadi_(small_johad)_in_village_Laporiya,_Rajasthan.jpg&oldid=557050918. Accessed 24 June 2021

WikiPanti (2011) Wikimedia commons contributors, 'File: Gadisar Lake, Jaisalamer.jpg', Wikimedia commons, the free media repository, 14 February 2021, 17:06 UTC. https://commons.wikimedia.org/w/index.php?title=File:Gadisar_Lake,_Jaisalamer.jpg&oldid=532808734. Accessed 14 February 2021

Wikipedia Contributors (2021) Rainwater harvesting. In: Wikipedia, The Free Encclopdia. https://en.wikipedia.org/w/index.php?title=Rainwater_harvesting&oldid=1029406823. Accessed 15 July 2021

Yannopoulos S, Antoniou G, Kaiafa-Saropoulou M, Angelakis AN (2017) Historical development of rainwater harvesting and use in Hellas: a preliminary review. J Water Sci Tech: Water Supply 17(4):1022–1034

Chapter 2
Classification of Rainwater Harvesting Systems

Abstract Rain being the only source of fresh water, harvesting rainwater gains its rightful place of importance for ensuring adequate water for the future growth of human habitation. Rainwater harvesting as we understand now is a technology that has been quite effective in meeting the water requirement in ancient times. As it has already been mentioned earlier, rainwater harvesting basically is providing for a system to collect rainwater for future use. Rainwater harvesting systems can be categorized in different ways. It can be adapted for different needs and also depending on the funds that are available. It can also be categorized on the size of the storage, the distance of travel, or on the various catchment type, etc.

Keywords Classification · Ancient technology · Micro-catchment · Macro-catchment · Floodwater harvesting · Domestic rainwater harvesting

Rain being the only source of fresh water, harvesting rainwater gains its rightful place of importance for ensuring adequate water for the future growth of human habitation. Rainwater harvesting as we understand now is a technology that has been quite effective in meeting the water requirement in the ancient time. To revive this ancient technology to suit the present day environment, it is of prime importance that one needs to try and broadly understand the different kinds of rain harvesting system that are in vogue in the present times.

Any rainwater harvesting system as we understand is essentially is made of three basic components. These according to Worm and Van Hattum (2006) are:

Catchment:
The surface which directly receives the rainfall.

Delivery system:
Delivery systems (gutters) to transport the water from the collection surface to the storage reservoir.

Storage:
Where the harvested water is stored for future use. Figure 2.1 is a conceptual figure created by me to describe the components of Rainwater harvesting system.

As it has already been mentioned earlier, rainwater harvesting basically is providing for a system to collect rainwater for future use. Very common examples

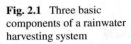

Fig. 2.1 Three basic components of a rainwater harvesting system

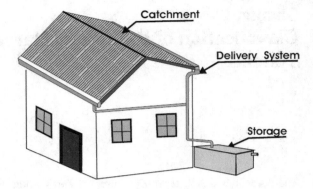

of which we see are the major dams and reservoirs that have been built for various purposes. This brings us to the classification of rainwater harvesting structures.

Rainwater harvesting can be classified in different ways, for e.g., based on the size of the structure, based on the use they have been designed for or what comprises the storage structure.

In 1991, the FAO of the United Nations, study "A Manual for the Design and Construction of Water Harvesting Schemes for Plant Production" by Critchley et al. divides rainwater harvesting into three major categories, classifying them according to the catchment area size and the runoff transfer distance.

The three categories were.

1. Internal or micro-catchment rainwater harvesting
2. External or macro-catchment rainwater harvesting
3. Flood water harvesting.

Other workers (Hatibu and Mahoo 1999; Mbilinyi et al. 2005; Ibraimo and Munguambe 2007; Mzirai and Tumbo 2010) modified this further and added an initial division to these three categories and called this "in -situ rainwater harvesting" or soil and water conservation, as its name suggests its function is to capture and store for rainfall directly in the soil, helping to increase soil infiltration and regeneration. Some prefer to exclude the third category of floodwater harvesting.

In their categorization of rainwater harvesting, Prinz and Malik (2002) suggest.

Internal or micro-catchment rainwater harvesting, also referred to as **within-field catchment systems**. All these are basically systems where rainfall is collected in small catchment areas ranging between 1 and 30 m (Critchley et al. 1991b).

This threshold of 30 m was increased by Oweis et al. (2001) to up to 1000 m². The peculiar characteristic of this system being that the runoff of these systems is stored directly in the soil, with no provision of overflow being provided for (Photo 2.1).

They cater directly to trees, bushes, or annual crops; good examples of these systems are the contour bunds, staggered counter trenches, etc. (Critchley and Siegert 1991; Dile et al. 2013; Falkenmark et al. 2001; Ibraimo and Munguambe 2007).

External or macro-catchment rainwater harvesting is also referred to as **long slope catchment technique**. Unlike the micro-catchment systems, they involve large

Photo 2.1 Internal or micro-catchment rainwater harvesting—continuous and staggered contour trenches near Supa Village, District Ahmednagar, on State Highway 27

catchment areas (30–200 m) to collect and have a provision of overflow of excess water.

Within this, the distance of travel to the targeted storage area is relatively large, according to Critchley and Siegert (1991), Falkenmark et al. (2001), and Ibraimo and Munguambe (2007) point toward this approach of the distance of travel being much more labor intense (Photo 2.2).

Another major difference between the micro- and macro-catchment rainwater harvestings is that the runoff capture is lower in the micro-catchment system as compared to what is collected in macro-catchment systems (Oweis et al. 2001).

Floodwater harvesting or **water spreading** or **spate irrigation** Oweis et al. (2001) preferred to group this together with external or macro-catchments systems because of the similar characteristics, such as the provision of overflow and the presence of turbulent runoff; however, their catchment area is far larger, covering several kilometers of distance (Critchley and Siegert 1991). Examples of this system are: permeable rock dams and water spreading bunds (Photo 2.3).

Oweis et al. (2001) added one more category and included a **domestic category**. To achieve this, they sub-categorized the micro-catchment systems, which include land catchment surfaces mentioned by the 1991 FAO study by Critchley et al. (1991b) added a non-land catchment surface, including rooftop systems, courtyards, and other impermeable structures. They justified this type of collection as being mainly used for domestic purposes. If the quality of the water is low, it could be also used in agriculture practices or to support home gardens.

Photo 2.2 Kaas Lake at Satara—an example showcasing a macro-catchment rainwater harvesting system

Photo 2.3 Floodwater harvesting—paddy fields near Pawan Village, District Pune

Barron (2009) classified rainwater harvesting based on the source of water (catchment area) into: in situ and ex situ technologies and manmade/impermeable surfaces. This division was founded on a proposal made by the SIWI. The common objective of both in situ and ex situ being reduction runoff water by enhancing soil infiltration (Barron 2009; Helmreich and Horn 2009; Mbilinyi et al. 2005). In this situation, water is collected directly where it falls and is stored in the soil (Cortesi et al. 2009). The ex situ technologies differ from the in situ systems and store runoff water externally to where it got captured (Barron 2009; Helmreich and Horn 2009).

Storage being essential after the rainfall, the rainwater needs to be collected; Barron (2009) provides a subcategory to divide rainwater harvesting in terms of the mode of storage. These systems can be located externally or underground. Some examples of this area: micro-dams, earth dams, farm ponds, sub-surfaces, sand dams or check dams, and tanks (Falkenmark et al. 2001).

According to Fewkes (2012), the storage capacity has relevant economical and operational implications for the system. When referring specifically to tanks, the material of construction—plastic, concrete, or steel—helps to determine its durability and cost.

Thus, rainwater harvesting can be subdivided based on source of water whether in situ or ex situ or no manmade/impermeable surface; this can be subdivided based on mode of storage, which being soil storage, within the subsurface, in a well or aquifer, and lastly within manmade structures such as dams, ponds, or tanks. This can be further subdivided based on the principal use the water is being put to, viz., forestry, crop production, livestock, lastly, domestic public or commercial use.

Finally, the term domestic rainwater harvesting (DRWH) system was formulated and has been used by authors such as Helmreich and Horn (2009). It is a category of rainwater harvesting that collects water for domestic purposes. It is mainly found in studies that analyze the spur of urbanization and how to cope with the rise in water demand in this area (Kahinda et al. 2007) (Fig. 2.2).

In short, rainwater harvesting systems can be categorized in all sorts of different ways. It can be adapted for different needs and also depending on the funds that are available. It can also be categorized on the size of the storage, the distance of travel, or on the various catchment type, etc.

In urban areas due to the space constraints, domestic rainwater harvesting design technique plays an important role. According to Fewkes (2012) out of the different methods currently used, the most common technology for collection is the rooftops. Rooftops in urban areas, especially so in the neo-metros, make excellent collectors of rainfall for domestic usage. To derive the full benefit from the rooftop systems, it is important to pay attention to the selection of construction material, slopes of roofs, maintenance, pollution, and extra water usage.

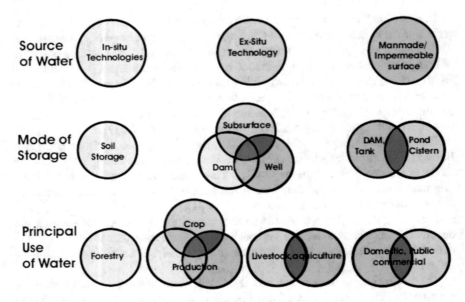

Fig. 2.2 Schematic of rainwater harvesting technologies based on source of water, storage mode, and principal use (Fox 2001)

According to Helmreich and Horn (2009) even though zinc and copper help to channel water easier than other systems, it is necessary to take into consideration the possibility of pollution due to the heavy concentration of metal. This fact actually makes rooftop harvested water unsafe for human consumption without treatment.

A typical rooftop terrace in the urban Indian cities in the present days is usually occupied almost totally by the solar water and/or electricity panels, which in the last decade have been made mandatory to provide for as per the new construction norms in India. This makes it extremely difficult to maintain a clean environment on the terraces of the buildings, due to which a building terrace is a host of contamination due to neglect. A study by Ibe and Ibe (2016), suggests that roof runoff water could impact negatively to the environment and if consumed without being treated may be injurious to human health (Photo 2.4).

The difficulty in maintaining a clean environment on the rooftops of such buildings which act as catchment areas will surely lead to the contamination of the rainwater that is being harvested, and hence, it is not advisable to use rooftop harvested water without prior treatment for drinking or even for other domestic use.

Photo 2.4 Showing solar panels installed on rooftop of "Tathasthu" building on Prabhat road, Pune

References

Barron J (ed) (2009) Rainwater harvesting: a lifeline for human wellbeing. United Nations Environment Program Nairobi, Kenya, p 69

Cortesi L, Prasad E, Abhiyan MP (2009) Rainwater harvesting for management of watershed ecosystems. In: Barron J (ed) Rainwater harvesting: a lifeline for human well-being. United Nations Environment Programme Institute, Kenya, pp 14–22

Critchley W, Siegert C (1991) Water harvesting manual, AGL/MISC/17/91. FAO, Rome

Critchley W, Siegert K, Chapman C (1991a) A manual for the design and construction of water harvesting schemes for plant production (AGL/MISC/17/91) FAO, United Nations, Rome, p 148

Critchley W, Siegert K, Chapman C, Finkel M (1991b) Water harvesting: a manual for the design and construction of water harvesting schemes for plant production. FAO, Rome, Rome, p 163

Dile Y, Karlberg L, Temesgen M, Rockström J (2013) The role of water harvesting to achieve sustainable agricultural intensification and resilience against water related shocks in sub-Saharan Africa. J Agriculture, Ecosyst Environ 181:69–79

Falkenmark M, Fox P, Persson G, Rockström J (2001) Water harvesting for upgrading of rainfed agriculture. Problem analysis and research needs. International Water Institute, Stockholm, p 94. https://siwi.org/downloads/Reports/Report2011WaterHarvestingforUpgrading2001.p94

Fewkes A (2012) A review of rainwater harvesting in the UK. Struct Surv 30(2):174–194

Fox P (2001) Supplemental irrigation and soil fertility management for yield gap reduction: On-farm experimentation in semi-arid Burkina Faso. Licentiate in Philosophy Thesis 2001:5 in Natural Resources Management. Department of Systems Ecology, Stockholm University, Sweden

Hatibu N, Mahoo H (1999) Rainwater harvesting technologies for agricultural production: a case for Dodoma, Tanzania. In: Kaumbutho PG, Simalenga TE (eds) Conservation village with animal traction, ATNESA. Harare, Zimbabwe, pp 161–171

Helmreich B, Horn H (2009) Opportunities in rainwater harvesting. In: WSIDDR International Workshop, Edinburgh, Scotland, UK, 28–30. Desalination; 248(1–3), Pub Elsevier B.V., pp 118–124

Ibe FC, Ibe BO (2016) Roof runoff water as source of pollution: a case study of some selected roofs in Orlu Metropolis, Imo state, Nigeria. Int Lett Natural Sci 50:53–61

Ibraimo N, Munguambe P (2007) Rainwater harvesting technologies for small scale rainfed agriculture in arid and semi-arid areas. Waternet Project, South Africa, PC17

Kahinda M, Jean-marc TA, Boroto J (2007) Domestic rain water harvesting to improve water supply in rural South Africa. J Phys Chem Earth Parts A/B/C(32):1050–1057

Mbilinyi BP, Tumbo SD, Mahoo HF, Senkondo EM, Hatibu N (2005) Indigenous knowledge as decision support tool in rainwater harvesting. J Phys Chem Earth 30:792–798

Mzirai O, Tumbo S (2010) Macro-catchment rainwater harvesting systems: challenges and opportunities to access runoff. J Animal Plant Sci, JAPS 7:789–800

Oweis T, Prinz D, Hachum A (2001) Water harvesting: indigenous knowledge for the future of the drier environments. ICARDA, Aleppo, Syria, p 40

Prinz D, Malik AH (2002) Runoff farming, WCA infoNet. Rome, Italy, p 39

Worm J, Van Huttum T (2006b) Rainwater for domestic use. Agromisa Foundation/Digigrafi, Wageningen, The Netherlands, p 82

Chapter 3
Rooftop Rainwater Harvesting

Abstract The difficulty in maintaining a clean environment on the rooftops of buildings which acts as catchment areas is a probable cause with results in the contamination of the rainwater that is being harvested, and hence, it is recommended that use of rooftop harvested water for drinking and even for domestic use should be done only after it has been subjected to prior treatment. A majority of government reports suggest that India is suffering from the worst water crisis in its history, and millions of lives and livelihoods are under threat. As per the records of the Ministry of Housing and Urban Affairs, there has been an increase in urban population from 29 crores in 2001 to 37.70 crores in 2011 which is approximately a 30% increase along with this the urban towns too have risen from 5161 to 7935 in the same period. This has resulted in an increase in the demand for water in the urban areas, at a time when these metropolitan cities in the country are already facing acute shortage of drinking water in the summers this despite the fact that these cities experience good rainfall and flooding water logging of streets in the monsoon period. The very fact that water harvested is water produced and thus the need to make sincere attempts to harvest every drop of water that falls within every premises, locality, city, and country.

Keywords Contamination of rainwater · Worst water crisis · Urban population · Acute shortage · Flooding · Waterlogging · Water management index · Decline in water level · Eco-friendly

The difficulty in maintaining a clean environment on the rooftops of buildings which acts as catchment areas is a probable cause with results in the contamination of the rainwater that is being harvested, and hence, it is recommended that use of rooftop harvested water for drinking and even for domestic use should be done only after it has been subjected to prior treatment. In India's context, the Standing Committee on Urban Development (2018–2019) cited from a Report of National Institution for Transforming India (NITI Aayog), "Composite Water Management Index—A tool for Water Management" by Kant (2018) suggesting that India is suffering from the worst water crisis in its history, and millions of lives and livelihoods are under threat. According to this report, nearly 600 million people in India face high to extreme water stress and about 200,000 people die every year due to inadequate access to safe water in the country. About three-fourths of the households in the country do not have drinking water at their premise. With nearly 70% of water being contaminated water, India is placed 120th among 122 countries in the water quality index.

The report also states that the crisis is only going to get worse, and by 2030, the country's water demand is projected to be twice the available supply, implying severe water scarcity for hundreds of millions of people; they also mention that National Commission for Integrated Water Resource Development of Ministry of Water Resources have also reported that the water requirement of the country by 2050 in high-use scenario is likely to be 1180 BCM, whereas the present-day availability is 695 BCM. The total availability of water possible in country is still much lower than the projected demand at 1137 BCM, which makes it important to further our understanding of our water resources and usage so as to be able to put in to place measures which would make our water use more sustainable.

As per the records of the MUD (2016), there has been an increase in urban population from 286.1 million in 2001 to 377.1 million in 2011 which is approximately a 31.14% increase along with this the urban towns too have risen from 5161 to 7935 in the same period. This has resulted in an increase in the demand for water in the urban areas, at a time when these metropolitan cities in the country are already facing acute shortage of drinking water in the summers. Despite the fact that these cities experience good rainfall and flooding water logging of streets in the monsoon period is a common yearly phenomenon during the rainy seasons.

Due to the increase demand for clean water, there is a dependency on the groundwater to fulfil this gap between demand and supply and on the other hand and the decreases in the quantum of natural recharge to the system due to urbanization has led to the depletion of groundwater levels. As per the CGWB, 2018 data, the premonsoon (April–May, 2018) water level data of wells shows that 66% of wells show a decline in water levels when compared with decadal average 2008–2017, it further states that certain prominent cities like Bengaluru, Chennai, Kolkata, Ranchi, Hyderabad, Vishakhapatnam, Mumbai, Guwahati, Indore, etc. have experienced steep fall in groundwater in terms of number of wells. Some prominent cities like Allahabad, Ghaziabad, Kanpur, Lucknow, Meerut, and Banaras have experienced up to a 100% fall in groundwater level.

The report of the National Institute for Transforming India (NITI Aayog) by Kumar (2019) mentions that most metro cities in India are water starved but not rain starved. The very fact that water harvested is water produced and thus we need to make sincere attempts to harvest every drop of water that falls within every premises, locality, city, and country.

Rain being the only source of clean water and is available in plenty in India, rainwater harvesting has been considered to be a simple, viable, and eco-friendly method of water conservation and a sustainable solution to recharge the groundwater (Kumar 2019).

As per CGWB, India 2000, "Guide to Artificial Recharge for Groundwater", the rooftop rainwater can be conserved and used for recharge of groundwater. The urban housing complexes or institutional buildings have large roof area and can be utilizing for harvesting rooftop rainwater to recharge aquifer in urban areas. Rainwater harvesting, in an urban setup due to constraint of open space for capturing the run off, is mostly achieved by subsoil storage by diverting the rainwater in to the subsurface aquifer systems during the rains (Raghavan 2009).

CGWB (2000) is of the opinion that the subsurface reservoirs are more attractive and technically feasible alternatives for storing surplus monsoon run off and that the subsurface reservoirs or aquifers are capable of storing substantial quantity of water.

The CGWB (2000), in their "Guide to Artificial Recharge for Groundwater", has presented a detailed tabulation of the availability of rainwater through rainwater harvesting depending upon the area of the rooftop and the annual rainfall in the particular area.

CGWB (2000) is also the view that the lithological suitability in conjunction with other considerations that are essential for creating subsurface storages is favorable the geological structures and physiographic units, of adequate dimensions and shape which cater to retention of substantial volume of water in porous and permeable formations.

According to CGWB (2000), subsurface reservoirs, which are located in suitable hydrogeological situations, are both environment friendly and economically viable. Any subsurface storage has advantages of not requiring flooding of large surface area, loss of cultivable land, displacement of local population, substantial evaporation losses, and sensitivity to earthquakes, nor is there any need to build massive structures to store water.

The storage of water in the subsurface would have a positive influence on the existing groundwater. The water levels could rise, resulting in a decrease in pumping costs. The injection of fresh water in to the aquifer would improve the quality of natural groundwater especially in areas with in brackish and saline groundwater.

The hydraulic connectivity of the aquifers would do away with the heavy cost of surface water conveyance system. This would further contribute to the increase in base flows of streams so as to positively impact the presently degraded ecosystem of riverine tracts. Last but not least, the structures required for recharging groundwater reservoirs are of small dimensions and cost, hence much cheaper as compared to the costs of check dams, percolation tanks, surface spreading basins, pits, etc.

As per the CGWB (2000), norms basic requirements to achieve artificial recharge are:

(i) Availability of monsoon runoff in space and time.
(ii) Identification of suitable hydrogeological environment and sites for creating subsurface reservoir through cost-effective artificial recharge techniques.

The most important aspect for achieving any groundwater recharge is the availability of source water. This can be evaluated by analyzing the monsoon rainfall pattern, its frequency, number of rainy days, and maximum rainfall in a day and its variation in space and time.

The next important aspect is the thorough understanding of geological and hydrological features of the area to determine type of recharge structure.

The aquifers absorb large quantities of water and do not release them too quickly, which means the aquifers having high vertical hydraulic conductivity and moderate to low horizontal hydraulic conductivity are best suited for achieving maximum artificial recharge. In natural environments, the combination of these two conditions is rare and estimating these parameters just from surface observation and lack of through understanding of hydrogeology is a formidable task and next to impossible. As a precautionary measure to ensure that there is no adverse effect due to the artificial recharge such as water logging, soil salinity, etc., the upper 3 m of the unsaturated zone is not considered for recharging.

Kumari (2009) believes that as there being a constraint of space for surface storage, and due to the falling water table, the water table is deep enough to accommodate additional recharge, which makes rainwater harvesting an ideal solution for solving water supply problems. The rainwater which is harvested can be stored in the subsurface storage to be used to meet the water demand on a later date. This view is also supported by CGWB (2017), in their manual on artificial recharge of groundwater.

Janhit foundations publication "*A People's Manual on Rainwater Harvesting*", by Hudda (1998), the Rainwater Harvesting Potential (RHP) or the amount of water that can be collected can be calculated based on the area of the catchment (A) in m^2 multiplied by the amount of annual rainfall (Rf)

$$RHP = A \times Rf$$

where
   A = Surface area of the catchment in m^2.
   Rf = Annual rainfall in m.

The total volume of water received to amount that can be actually harvested is dependent on the runoff coefficient of the roof surface, different slopes, and different roofing material have different runoff coefficients. In urban areas, it is usually concrete having a runoff coefficient of 0.9, but different slopes and different roofing material have different runoff coefficients.

This makes it very easy to calculate the amount of water that can be harvested annually, the actual amount of water that is available is about 80% of the harvesting potential, as one needs to take in account the evaporation and transmission loss and the times when the precipitation is very low and does not generate any surface flow.

This is the simple part where rainwater harvesting is concerned, and hence, it has led to many people undertaking the design and implementation of rooftop rainwater harvesting systems which was made mandatory for any construction in the urban cities in India.

Most rooftop harvesting systems in urban centers are focused on groundwater recharge due to the constraint of cheap space and secondly due to the number of rain days being limited and the gap between the rain and non-rainy days being quite large diverting rooftop water in to storage tanks is not advised, especially when there is availability of fresh clean groundwater to supplement the daily water requirements (CGWB 2000).

The common practice that is usually adapted by most practitioner's is to pass the rooftop harvested water through some kind of filtration system and divert it in the ground via pits, dug-wells or drilled borewells, or in to specially constructed tanks. There are plenty of designs for rooftop rainwater harvesting systems which cater to recharging groundwater, these are available in the form of manuals from state and central government and non-government organizations or on their websites. The CGWB (2003) manual *"Rain water Harvesting Techniques To Augment Ground Water"* describes various techniques of augmenting groundwater by various structures, viz. recharge pit, recharge trench, an existing tube-well, with an intermediary filtration tank or an online filter. In case of multi-storied building with numerous down-take pipes area connected to a single filtration tank, here the tank is provided with multiple recharge tube-wells, based on the recharge capacity of the aquifer.

The KSCST website even shows how one can achieve groundwater recharge by using homemade filters by reusing barrels instead of constructing a interemediary tank. The various filtration aids that are commonly used to filter the rooftop water before it is injected in to the groundwater as suggested by CGWB (2003) are the sand bed filter which consists of a constructed tank with layers of pebbles at the base followed by a layer of aggregate and on top of it a layer of sand the rooftop water is diverted in to the top of the tank and it gets filtered as it passes downwards through these layers by gravity and emerges out at the base of the tank and this is then diverted in to a storage tank for collection for future use.

This kind of filtration system involves civil work and some space above ground to construct the sand bed filter. Where there is a constraint of space, they have a pop up filter that can be installed and can be connected to the storage tank. In situations where there is a large volume of rainwater, they recommend a stabilizing tank with baffle walls dividing it in to 4 separate chambers which are connected at different levels to ensure that only clear water manages to flow out and the suspended matter (if any) gets deposited within the first two chambers itself.

These and many more of such harvesting designs freely available on the internet; this in the last two decades has encouraged a lot of new entrepreneurs to venture in this new lucrative field. Some of the rainwater harvesting structures that are being executed and are in vogue the other filters which are constructed and have been observed at various locations in and around Pune city have been shown below (Photos 3.1, 3.2, 3.3, 3.4, and 3.5).

The CGWA (2021) website also suggests Standard Design of Rooftop Rainwater Harvesting Structure and Recharge for Groundwater Abstraction up to 10 m^3/day; this is applicable for Delhi and environs, it basically incorporates a tube-well drilled down to about 15 m with slotted casing at the base to facilitate groundwater recharge.

CGWA (2021) suggests that the depth of the recharge well should be determined based on post-monsoon depth to water level and should be kept 2–3 m above post-monsoon water level. They claim their design to be indicative and would vary as per site condition. In case of industries such as chemical, pesticide, slaughter house, textile, dyeing, etc., which are likely to contaminate groundwater it is advised to collect rooftop rain water and store the same for reuse.

Photo 3.1 Indigenously designed Barrel filter for groundwater recharge of rooftop water at GURUKRUP Apartment Lane no 15, Prabhat road Pune

Apart from these, there are of the shelf commercial filters available in the market which can be installed inline on the down-take pipes to filter the rooftop water before diverting it to the recharge well or borewell. Some of which have been shown below (Photos 3.6, 3.7, and 3.8).

These commercially manufactured and many more of the indigenously designed filtration systems have been installed by various individuals and agencies to try and implement the rainwater harvesting systems which have been made mandatory since 2005.

There have been scientific correspondences which recommend converting dried-up borewells as groundwater recharge wells (Ramachandrula 2018). This is possible, as the borewell probably have dried up due inadequate recharge or over exploitation, hence such borewells which in the past were able to provide substantial quantities of water, which have noticed a decline in yield over the years could be subject to recharge and they tend to yield good results. In contrast to this, those borewells which had not encountered any water while drilling when subjected to rainwater harvesting tend to yield water only during the monsoon season at a time they are being constantly being recharged with rainwater, but then to dry out within days of the end of the monsoon season.

Photo 3.2 Custom-made rainwater harvesting online filter installed at Visava Heights, Aundh, Pune

Photo 3.3 Fabricated rainwater harvesting filter installed at Visava Gharkool Society, Aundh, Pune

A vast majority of the agencies involved in designing and installing rainwater harvesting systems are usually headed by social activist, architects, engineers, and entrepreneurs, who have limited or zero knowledge of hydrogeology, and it has become a common practice to go by what they are able to gather online or through

books and what is being practiced in the west without have a clue regarding the
hydrogeological constraints that one faces within hard rock terrains, which constitutes a vast majority of the area in India and hence one can say that they tend to
blindly follow available designs while implementing rooftop rainwater harvesting
projects.

Figure 3.1 depicts a typical standard generalized rainwater harvesting system
generally adapted by most workers in the field.

These designs usually include a considerable amount of storage for the rainwater
before it gets diverted for recharge. This probably an outfall due to the need for
the project to get certified as a Green Building by IGBC, GRIHA or other rating
agencies. The basic mandatory requirement of a project getting accredited as Green
Building is when provisions of adequate storage within the aquifer or within storage
tanks or within the rainwater harvesting structure have been provided for.

In absence of adequate knowledge regarding the aquifer storage which is lacking
even in many of the hardcore geologists who after retirement found this an lucrative
business and hence have landed up designing the rainwater harvesting systems with
emphasis being given to storage tanks and/or provision of large buffer tanks within

Photo 3.5 Recharge
chamber with borewell with
rock aggregate and charcoal
mixture as filter material
–Kohinoor Falcon Site, Sus
village, Pune district

rainwater harvesting structures instead of trying to ascertain the recharge capacity of the aquifers, which is more important when it comes to arresting groundwater levels from declining due to inadequate recharge.

According to the IGBC requirement, the design of rainwater harvesting system should be such as to capture at least "one-day rainfall" from roof and non-roof areas. While designing something based on such requirement requires either a very high recharge capacity, or a large number of recharge structure with large buffer tanks or a very large storage tank for storing rainwater. The quantity of "one-day rainfall" that needs to be harvested is estimated based on the suggested percent of average peak month rainfall; this is the month during which the area received the maximum precipitation, where the average peak month rainfall is up to 250 mm, the one-day rainfall assumed is 9% of the average peak monthly rainfall, which is the lower end and where the value is between 501 and 700 it is 4.5%, which is the higher side of amount of annual rainfall for most parts of India.

To arrive at the average peak month rainfall, an average of at least last 5 years peak month rainfall of the project location is considered (Table 3.1).

The above value for average rainfall based on the above calculations for Pune district results to about 29.58 mm which is slightly higher than 1.5 times the average daily rainfall for Pune, which is only around 18 mm. The criteria for getting extra

Photo 3.6 RAINTAP FILTER, installed at Institute of Management Development & Research, Deccan Education Society Campus, Agarkar Road, Pune, Maharashtra

points related to the rainwater harvesting that needs implemented based on the quantity of water that constitutes "one-day rainfall" is even higher.

As per IGBC (2019) for areas with rainfall up to 250 mm to gain 1 point the value to be considered for one-day rainfall is 12% of average peak month rainfall to gain the full 4 points it is as high as 21%.

In case of project sites within Pune district having annual rainfall 701 mm and above, to be awarded 1 point the value to be considered for one-day rainfall is 4% of average peak month rainfall and to be awarded the full 4 points; the value to be considered for one-day rainfall is 7% of average peak month rainfall (Table 3.2).

As per these criteria, for Pune this value to qualify for 4 points it is nearly 47.34 mm rainfall needs to be considered while calculating the one day's runoff, which is more than two and half the time of the actual average precipitation. Here the value for "one-day rainfall" that is considered is just short of the peak intensity rainfall value for that particular area; this would result in runoff being only during cloud bursts. This situation in itself is detrimental to environmental sustainability, as it would result in a surface runoff which is far below the original and would thereby go to affect the river ecology of the area and subsequently affect the others in the downstream region too.

To be rated under the IGBC exemplary performance under ID Credit 1 which is Innovation in Design Process, if rainwater run-off from roof & non-roof areas is

Photo 3.7 Online filter "RAINWAY", installed at Narshima Society, Boatclub Road, Pune

captured and/ or recharged, the percentage of rainwater that needs to be considered is given in Table 3.3.

To cater to this amount of recharge capacity would result in a system which is highly overrated and would really be functioning with adequate capacity during the times of cloud bursts and probably would be resulting in zero surface runoff.

There is a special provision for projects where in the static water level within the project sites are high, as per IGBC (2019). If the water table is less than 8 m, here again no consideration is being given for the difference between the static water level which is attained in wells tapping the unconfined aquifer system, and the piezometric levels in wells tapping the confined or semi-confined aquifer system. In such cases, the projects, to be eligible for additional points it is mandatory for the project to provide rainwater storage systems to capture the runoff volumes as per Table 3.4.

The above-mentioned requirements are contradictory to the CGWB norms regarding construction of artificial storage tanks in areas where rainfall is limited to a few months. The Guide to Artificial Recharge for Groundwater (CGWB 2000) and the Rainwater harvesting Manual by Central Public Works Department (CPWD 2002), clearly mentions that where the gap between rainy days and non-rainy days is

Photo 3.8 Multiple online filters "RAINWAY", installed on roof down-take pipes at Blessed Sacrament Scholasticate Site, Vadgaonsheri, Pune

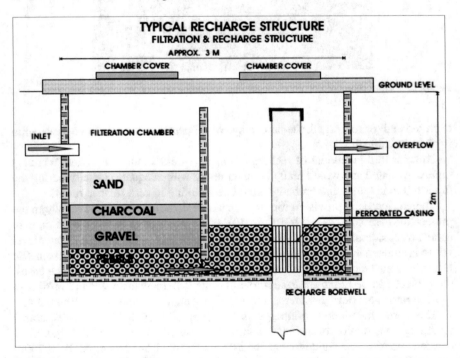

Fig. 3.1 Typical recharge structure of recharge borewell in chamber with filter (Modified after Pandit et al. 2018)

Table 3.1 Mandatory criteria to arrive at "one-day rainfall" (After IGBC 2019)

Sl. no	Average peak month rainfall (in mm)	One-day rainfall (% of average peak month rainfall)
1	Up to 250	9
2	251–350	7.5
3	351–500	6
4	501–700	4.5

Note To estimate the average peak month rainfall, consider the average of at least last 5 years peak month rainfall of the project location

Table 3.2 Criteria to arrive at "one-day rainfall" for additional points (After IGBC 2019)

Sl. no	Average peak month rainfall (in mm)	One-day rainfall (% of average peak month rainfall)			
		1 point	2 points	3 points	4 points
1	Up to 250	12	15	18	21
2	251–350	10	12.5	15	17.5
3	351–500	8	10	12	14
4	501–700	6	7.5	9	10.5
5	701 & above	4	5	6	7

Note To estimate the average peak month rainfall, consider the average of at least last 5 years peak month rainfall of the project location

Table 3.3 Criteria to arrive at "one-day rainfall" for exemplary performance (After IGBC 2019)

S. no	Average peak month rainfall (in mm)	One-day rainfall (% of average peak month rainfall)	
		Case A	Case B
1	Up to 250	24	18
2	251–350	20	15
3	351–500	16	12
4	501–700	12	9
5	701 & above	8	6

Note To estimate the average peak month rainfall, consider the average of at least last 5 years peak month rainfall of the project location

more, then creating a storage tank for collection of rainwater is not advisable, unless the groundwater in the area is saline or not available, and BIS (2016), the National Building Code of India recommends that the CGWB norms be followed where it comes to rainwater harvesting and recharging the groundwater systems.

Moreover in high-rise buildings, usually one-day storage is near about or less than one-day fresh water requirement, which means that the storage tank that is

Table 3.4 In projects with groundwater levels less than 8 m, criteria to arrive at "one-day rainfall" (After IGBC 2019)

Sl. no	Average peak month rainfall (in mm)	One-day rainfall (% of average peak month rainfall)			
		1 point	2 points	3 points	4 points
1	Up to 250	6	9	12	15
2	251–350	5	7.5	105	12.5
3	351–500	4	6	8	10
4	501–700	3	4.5	6	7.5
5	701 & above	2	3	4	5

Note To estimate the average peak month rainfall, consider the average of at least last 5 years peak month rainfall of the project location

constructed would probably never fill up to its full capacity, and would be empty for nearly 300 days in a year, which means it becomes a liability from maintenance point of view as well as ensuring that the water that is collected within it is contaminant free.

Many a times, the rainwater harvesting consultants do not suggest a surface storage tank, instead the rainwater harvesting structures are so designed that it incorporates a large capacity buffer tank of capacity of nearly 10–12 m^3 or more, and based on the requirement for "one-day water", the required number of rainwater structures is put in place, with little or no regards to the prevailing hydrogeological conditions.

Technically speaking this is done so that the rainwater harvesting system which is put in place is within the preview of the Central Groundwater Board norms and it does not really violate the Central Groundwater Board clause of no storage of rainwater in most parts of India where the rainfall is limited to only 4 months of the rainy season in a year.

Naturally, when such rainwater harvesting system designs get implemented without really giving any thought to the hydrogeology and without trying to understand the limitations of each design and the impact it could have when implemented in the wrong setup, and instead of this being reprimanded as bad practice, it is lauded and awarded extra points.

Ideally speaking, designing of rainwater harvesting systems should be to cater to the average rainfall only so that there is some runoff generate during cloud bursts, etc. To maintain the environmental and ecological balance there by only the amount of increase in runoff should be catered for while designing the rainwater harvesting system and may be about 10% more as the groundwater is being abstracted for supplementing the daily water requirement.

Naturally in light of the fact that such haphazardly implemented rainwater systems which is primarily undertaken to gain additional points in the Green Building rating system eventually lands up doing more harm than good, and over the last two decade many such systems have being implemented, and it really has not benefitted in terms of meeting the water requirements, nor has it helped in stop the declining the groundwater levels, on the contrary it has led to the contamination of the aquifers in general, and causing water to stagnate in the chambers and creating an ideal environment for the breeding of disease spreading mosquitoes.

This is the main reason why such haphazardly implemented systems are subject to neglect and become defunct in the coming years. The best part is that none of the rainwater harvesting system designs, that are freely available, really specify the Standard Procedures for maintenance, nor do they specify any maintenance schedules, nor is there any reference to the geological and hydrogeological setup they are suited for, and eventually leads to most of these systems also getting defunct within a short time after being implemented, as they usually do not manage to cater to the recharge of groundwater they had been designed for.

Within the state of Maharashtra, as per the Government Directives of Maharashtra Regional & Town planning Act 1966, since 2017, a minimum of 1 recharge structure per 5000 m^2 of built-up or storage. It is a good initiative, but lacks sound technical backing, as it does not take into account the redundancy of constructing a surface storage structure nor does it take in to account the clause of the depth of water table which is very essential to ensure that there is no upward movement of salts due to water logging, etc., leading to the risk of contamination to the aquifers.

More over with the construction norms which allow for high-rise buildings the total built-up area within a plot is very high and many a times due to the permissible excavated area being just 1 m from the plot boundaries and no mandatory requirement for having a mandatory Green area to be on natural ground, makes it difficult to really manage to find adequate distance between the recharge structures that need to be implemented and almost all the rooftop or storm water chambers need to be connected to some sort of an recharge structure.

All these systems described earlier are basically designed to harvest rainwater, which is assumed to be free of contaminants, but in reality not just the surface runoff but even rooftop waters have a host of contaminants, namely zinc, copper, and lead (Simmons et al. 2001), volatile organohalogen compounds, petroleum hydrocarbons, various ions, as well as various pesticides (Polkowska et al. 2002). There is a common occurrence of various pathogenic microorganisms in samples taken from rainwater systems (Lye 2002).

Shrivastava et al. (2019), too are of the opinion that there is a possibility of development of microbes in stored water, making it unsafe for drinking. Abbassi and Abbassi (2011), have attempted to trace the pathways by which pollutants can enter in a rainwater harvesting system, they have tried to come up with strategies to manage the water quality at preharvest as well as post-harvest stages.

We at Well & Water Works over the years have experimented with various designs and have come up with a design of our own which we feel is a very practical and easy to implement, and has a low cost. It is a 3 chamber filtration system as shown in Fig. 3.2 that can be implemented to achieve groundwater recharge; this system is not restricted for harvesting only the rooftop but is designed to harvest surface runoff too.

The first chamber is basically a modification of an existing storm water chamber, and it incorporates an oil trap and settling chamber. The second chamber is the filter

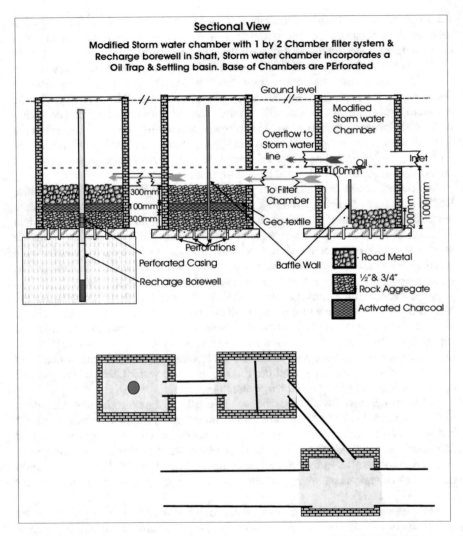

Fig. 3.2 A three chamber filtration system for groundwater recharge, Well & Water Works, Pune

chamber, having a gravity and reverse gravity filtration with a layer of activated charcoal which would help in absorption of some of the unwanted smell and hydrocarbon impurities. The third chamber is provided with a recharge borewell, the depth of which is dependent on the depth at which the aquifers are encountered at the time of drilling, but currently limited to a maximum of 60 m due to the Government of Maharashtra restriction of drilling beyond 60 m. The total numbers of recharge structures that need to be implemented are based on the recharge capacity of the aquifers that are encountered while drilling the recharge borewells.

The chamber with the recharge borewell too has a gravity filter system, through which the water passes before being diverted into the aquifer via the borewell. It is different from other systems, as the overflow that is generated is, before filtration hence chances of the filter getting choked and the catchment area getting flooded is minimized.

For ease of maintenance and easy access for cleaning purpose, the dimensions have been kept to a maximum of 1 m × 1 m × 1 m below inlet level for modified storm water chamber and the filter chamber. The depth of the recharge chamber below the inlet level can be adjusted according to the rainfall intensity and the recharge capacity of the borewell to minimize overflow from the system during cloud bursts.

The chambers in this system are provided with a perforated base to facilitate percolation of remnant waters into the subsurface. This in turn minimizes the chances of microorganism growth in the system due to water stagnation, which usually is the case with some of other indigenously designed filtration systems. The maintenance suggested is once a year immediately after the first premonsoon showers, during which the filter media usually tend to get clogged due to high quantity of suspended matter in the flowing water.

All over the world, one can find rainwater harvesting being successfully implemented to meet their daily water requirements. A study by Han & Andrews (2017), in their studies in Singapore and Vietnam, revealed that, in Singapore, a city-state where fresh water resources is limited harvesting rainwater was a natural extension of pre-existing strategies to reduce, reuse, and replenish water sources. A vast majority of the population of Singapore, i.e., 86% lives in high-rise buildings; hence, rooftop water collection systems have been installed to maximize the use of rainwater. The rainwater is collected in tanks which is then used for toilet flushing, thereby helping in reduction in water consumption, saving on energy and other costs within the buildings.

Similarly in a village near Hanoi, Vietnam, without piped water supply, and the groundwater being contaminated with arsenic and the river water too is polluted, and bottled water being too expensive, leaving them with one option for drinking water that is to use rainwater. Here, several community-based rainwater harvesting systems, including within public schools and hospitals, are successfully supplying drinking water to residents.

Jordan is one of the world's four poorest nations in water resources. The findings of Awawdeh et al. (2011) indicate that a maximum of 99,000 m³/year of rainwater can be collected, of which 37,000 m³/year of it from roofs of buildings and balance from open impervious areas, provided that all surfaces are used and all runoff from the surfaces are collected. This would lead to a saving of nearly 125–145% of the total potable water.

In their review paper on Roof harvested Australia, Chubaka et al. (2018), it can be concluded that a mandatory rainwater tank plumbing policy on houses in new developments was enforced in South Australia, Victoria, New South Wales and Queensland where in having a rainwater tank on large extensions became mandatory, but due to the lower quality of the harvested water in comparison to that being supplied by municipalities, the Australian Federal Government and all States Health Departments recommend the public to exclusively limit rainwater use for non-potable purposes to avoid risks of contamination.

Many such cases from other countries both developed and developing countries some even from Southern Parts of India can be cited. These case studies are located within differing climatic and geologic setups, and where the quality of groundwater that can be tapped is not good and hence, storage tanks need to be an integral part of the rainwater harvesting system and maintenance of the quality of the harvested water being a crucial component, and it is helping in saving a substantial quantity of potable water and reducing the load on the municipal supplies.

In contrast to this in due to the constraint of space in the urban metros situated within the Hard rock terrain of Deccan basalts due to prevailing the climatic, viz. limited no of rainy days along with large dry spells between rainy days along with prevailing the geologic conditions, storage tanks are not recommended, instead recharge to the groundwater is what is being recommended as a means of harvesting and storage of rainwater.

In spite of all the government initiative and legislation related to rainwater harvesting to meet the shortfall of water in the summer months and also to recharge the groundwater aquifers for more than a decade has done little in terms of overcoming the water woes of the people at large, come summer there is always news of falling water table and people facing water shortages as their well go dry. The CGWB, under the Ministry of Water Resources, published a report in 2017 which found that the groundwater level in India has depleted by 61% from 2007 to 2017.

With the onset of summer, it is common for the daily newspapers and television news channels being flooded with stories highlighting the grime situation of water scarcity faced by metro cities in India. Some of the articles in news which highlight this have been shared below: "Nearly 76 million people in India do not have access to safe drinking water, as polluted rivers and poor storage infrastructure over the years has created a water deficit which may become unmanageable in the future." (NDTV Ltd. 2017).

"As Chennai struggles with unprecedented levels of water scarcity and the city on a list of those likely to run out of groundwater by 2020, residents are graduating From harvesting rain water to recharge the water table to using rain water directly" (NDTV Ltd. 2019).

In an article in the Asian Age, by Telang (2018), it posts a very grim picture of groundwater within the state of Maharashtra, according to it in a report submitted in the Lok Sabha in March. Nearly 57% of wells in Maharashtra have shown a decline in groundwater levels.

The above articles and countless other such reports in daily newspapers and TV channels across India which usually seem to show up during the summer months is a clear indication that rainwater harvesting as it is being implemented by various government and non-government agencies as well as private individuals do not seem to be really helping, this is especially true within the Deccan basaltic terrain.

To really understand this, why in spite of so much of effort being put in by the government by way of legislation and it is implementation and discounts on property tax; the situation does not seem to be improving is something of grave concern. The news of water level declining is even more disturbing in the hard rock basaltic terrain of India which is prone to regular droughts due to failure of the monsoons. In my opinion, it is very essential that one needs to have a real picture of the hydrogeological setup within Deccan basalts, into which the harvested water is being diverted. This, obviously is lacking and hence what is being done by most players currently active in this field is in a way technically incorrect, and there is an urgent need for this to be rectified and to do this one needs to understand the Deccan basaltic aquifer system.

References

Abbasi T, Abbasi SA (2011) Sources of pollution in rooftop rainwater harvesting systems and their control. Critical Rev Environ Sci Tech 41(23):2097–2167

Awawdeh M, Al-Shraideh S, Al-Qudah K, Jaradat R (2011) Rainwater harvesting assessment for a small size urban area in Jordan. Int J Water Res Environ Eng 4(12):415–422

BIS (2016) National building code of India, vol 2. Bureau of Indian Standards, Manak Bhavan, New Delhi, India, p 98

Central Groundwater Authority (2021). http://cgwa-noc.gov.in/LandingPage/GuidelinesonlineFilling/StandardDesignRooftopRain.pdf. Accessed 10 May 2021

Central Public Works Department (2002) "Rainwater harvesting- Manual", Government of India, Nirman Bhavan, New Delhi110011 pp 83

CGWB (2000) Guide to artificial recharge for groundwater. Ministry of Water Resources, New Delhi, India, p 75

CGWB (2003) Rain water harvesting techniques to augment ground water. Ministry of Water Resources, Faridabad, WB, India

CGWB (2017) Report of the ground water estimation committee, GEC-2015. Ministry of Water Resources, River Development & Ganga Rejuvenation Government of India, New Delhi, India

Chubaka CE, Whiley H, Edwards JW, Ross KE (2018) A review of roof harvested rainwater in Australia, Hindawi. J Environ Public Health 2018(6471324), 14 p. https://doi.org/10.1155/2018/6471324

Government of Maharashtra Directives (2017) No.TPS-1816/CR-443/16/DP Directives/ UD-13, 13 April, Under Section 154(1) of Maharashtra Regional & Town planning Act 1966

Han M, Andrews L (2017) International Water Association website https://iwa-network.org/can-rainwaterharvesting-transform-cities-into-water-wise-cities/. Accessed 30 December 2021

Huddah (1998) A people's manual on water quality. Janhit Foundation Meerut, p 15

IGBC (2019) IGBC green homes rating system—Version 3.0 for multi-dwelling residential units. Abridged Reference Guide, IGBC, Hyderabad, India

Kant A (2018) Composite water resources management—performance of states. National Institute for Transforming India (NITI Aayog), Government of India, New Delhi, Report p 179

Kumar R (2019) Composite water management index. NITI Aayog in association with Ministry of Jal Shakti and Ministry of Rural Development, Government of India

Kumari P (2009) Design & policy issues on rainwater harvesting in India. Dissertation submitted to National Law School of India University Bangalore in partial fulfilment of Postgraduate Diploma in Environment Law, Bangalore, India

Lye DJ (2002) Health risks associated with consumption of untreated water from household roof catchment systems. J American Water Res Assoc 38:1301–1306

Ministry of Urban Development (2016) Handbook of urban statistics. Government of India, New Delhi, p 383

NDTV Ltd. (2017) https://swachhindia.ndtv.com/76-million-don't-have-safe-drinking-water-indias-looming-water-crisis-5606/. Article dated September 22. Accessed 1 May 2021

NDTV Ltd. (2019) https://www.ndtv.com/cities/chennai-water-crisis-chennai-residents-use-rain-water-harvesting-to-get-25-000-litres-of-rain-water-2058690. Article dated 25 June. Accessed 1 May 2021

Pandit K, Lalwani K, Lalwani A (2018) Practical issues related to effective rainwater harvesting within The Deccan Basaltic Province with special reference to the hard rock areas of Pune & Environs, Abstract Presented. G. D. Bendale Memorial National Conference on, Ground Water: Status, Challenges and Mitigation. JALGAON, Maharashtra, India

Polkowska Z, Górecki T, Namieśnik J (2002) Quality of roof runoff waters from an urban region (Gdańsk, Poland). Chemosphere 49(10):1275–1283

Raghavan S (200) Rainwater harvesting in India: traditional and contemporary. https://www.ind iawaterportal.org/articles/rainwater-harvesting-india-traditional-and-contemporary. Accessed 1 May 2021

Ramachandrula VR (2018) Scientific correspondence. Curr Sci 115(2):25

Shrivastava PK, Patel D, Nayak D, Satasiya KF (2019) Harvesting and potable use of rooftop rain water to tackle imminent drinking water crisis in Coastal Gujarat, India. Current J Appl Sci Tech 35:1–10

Simmons G, Hope V, Lewis G, Whitmore J, Gao W (2001) Contamination of potable roof-collected rainwater in Auckland, New Zealand. Water Res 35(6):1518–1524

Standing Committee on Urban Development (2018–2019) (2019) Rainwater harvesting in metropolitan cities; 16th Lok Sabha Ministry of Housing and Urban Affairs, 24th Report, Lok Sabha Secretariat, New Delhi

Telang S (2018) https://www.asianage.com/byline/sonali-telang. Accessed 1 May 2021

Chapter 4
Rainwater Harvesting and The Deccan Basalts

Abstract Rainwater harvesting within the Deccan basaltic terrain is not really as simple as it is assumed to be. As mentioned in the earlier chapter, come summer the news articles in daily newspapers, regarding failed borewells, falling water table, insufficient water from borewells in major cities within the Deccan basaltic province is of common occurrence. This is not surprising, as the Deccan basalts are one of the most enigmatic rocks and pose a very difficult task where it comes to groundwater exploration and naturally also for groundwater recharge, especially in the urban areas where due to constraint of space one needs to resort to borewells for recharge purpose.

Keywords Basalts · Hard rock terrain · Multi-layered aquifer system · Inter unit zone · Compact basalt flow unit · Amygdaloidal basalt flow unit · Sheet joints · Red horizon · Down-the-hole drilling

In spite of all the efforts being put in by the government by way of legislation and it is implementation and discounts on property tax etc. has really not helped in improving the situation and the news of declining water levels is a cause of concern especially within the hard rock basaltic terrain of India which is prone to regular droughts due to failure of the monsoons.

This makes it essential that one has the right information regarding the hydrogeological setup within Deccan basalts, into which the harvested water is being diverted. What is obvious and amply clear from what has been discussed the preceding chapters is that there is an obvious lack of understanding of the hydrogeological system prevalent as far as Deccan basalts are concerned and what is being done by most players currently active in this field, as far as rainwater harvesting is concerned currently is in a way technically incorrect, which in turn highlights the urgent need for this to be rectified and to do this one needs to really understand the Deccan basaltic aquifer system.

Rainwater harvesting within the Deccan basaltic terrain is not really as simple as it is assumed to be. As mentioned in the earlier chapter, come summer the news articles in daily newspapers, regarding failed borewells, falling water table, insufficient water from borewells in major cities within the Deccan basaltic province is of common occurrence. This is not surprising, as the Deccan basalts are one of the most enigmatic rocks and pose a very difficult task where it comes to groundwater exploration and

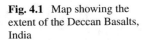

Fig. 4.1 Map showing the extent of the Deccan Basalts, India

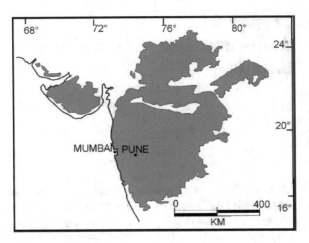

naturally also for groundwater recharge, especially in the urban areas where due to constraint of space one needs to resort to borewells for recharge purpose. Nearly two-thirds of the western, central, and southern peninsula of India are made up of different hard rocks such as basalts, granites, gneisses, etc. Most workers tend to group the basalts along with the other hard rocks where hydrogeology is concerned with an assumption that the fractures and joints give rise to the aquifer systems within it. In reality, the basalts, even though they fall under the category of hard rocks, the aquifer systems developed within them are in no way similar to those that occur within the granitic and other hard rock terrains. Deccan traps, unlike other hard rocks, form a multi-aquifer system (Rai et al. 2011).

The Deccan basalts or The Deccan traps as they are more popularly referred to occupy vast parts of the Western and Central India lying between 69° and 79° E long. And 16°–22° N lat. constitutes one of the largest volcanic provinces on the earth. (Fig. 4.1) Deccan traps sequence comprising of multiple layers of solidified lava flows is more than 2000 m thick on its western margin near the coast of Mumbai, and its thickness decreases eastward, on its eastern margin which is located west of Nagpur city it is only 50–100 m in thickness; the total area covered by the Deccan basalts is ~500,000 km^2 area which is spread over parts of the states of Maharashtra, Madhya Pradesh, Gujarat, Chhattisgarh, Goa, Northern Parts of Karnataka and Andhra Pradesh, and South Eastern parts of Rajasthan (Photo 4.1).

The occurrence of ground water being limited in quantity within hard rock terrains such as Deccan traps is a well-known and well-documented fact. The type of basalt, its degree of weathering, the nature and intensity of fracturing control the occurrence and movement of ground water resources in the Deccan basalt (Kulkarni et al. 2000). The main aquifer horizons occur in vesicular, fractured, and weathered amygdaloidal basalts (Pawar 1995). According to Pathak et al. (1999), the aquifers in hard rocks have limited storage. Individual flows comprised of an upper vesicular unit and a lower massive unit which may or may not be fractured/jointed. A sedimentary

Photo 4.1 Typical Deccan Basaltic topography showing multi-layered basalt flows seen at Varandha Ghat, District Raigad, Maharashtra

Intertrappean Bed may at times separate two lava flows, and hence, they give rise to confining conditions.

The fracture zones act as conduits that facilitate infiltration to the deeper zones (Deolankar et al. 1980). According to Kale and Kulkarni (1992), fracture lineaments have been proven to serve as conduits for groundwater infiltration and transmission and are important from the recharge and exploitation point of view.

According to Singhal (2009), the Deccan trap consists of as many as 32 lava flow units. The fractured and weathered basalts are the ones that contribute to the main aquifers, at places the intertrap formations also form aquifer horizons. The transmissivity ranges between 1 m day and about 500 m day. The intensity of fracturing rather than the age of the basalts is what determines the well yields.

According to Kulkarni et al. (2000), the key to hydrogeology of the basalts lies within the understanding of the physical framework within which the groundwater moves within these rocks. They further state that the vesicular amygdaloidal basalt and the compact basalt are the two types of basaltic flows which occur as alternate layers in the volcanic pile, and it is the structures such as sheet joints and vertical joints within the generally inhomogeneous basalt rocks that serve as zones of groundwater flow.

Varade et al. (2017) in their research findings also emphasize on the role of lineament mapping in hard rock aquifer system for identification of groundwater potential zones.

Down-the-hole hammer (DTH) is being the most commonly used method for drilling wells within the Deccan basalt, and it is also the cheapest as well as the

Photo 4.2 Common site of dust blowing while borewells are being drilled within the Deccan basaltic terrain using the down-the-hole hammer drilling rig

fastest available method for the drilling of borewells within the hard rock terrain of the Deccan basalts in Maharashtra (Photos 4.2 and 4.3).

Duraiswami et al. (2012) state that the basaltic flows which constitute the shallow and deeper aquifers within the Deccan basalts are hydrogeologically speaking very inconsistent and complex, and hence, the potential of these aquifers is highly variable and inconsistent when it comes to the ascertaining sustainability of the source.

According to Singhal (1997) and Lalwani (2004), the storage volume within the hard rock aquifers of the Deccan basalts is limited and the aquifers normally are recharged to full capacity during monsoons season where the region experiences normal rainfall. In spite of this, due to the over exploitation of the system the borewells tend to have a remarkable decrease in the yield that can be derived from them, and this leads to water scarcity and shortage especially in the summer months of April and May.

In the early days, during the late 70s and early 80s, the shallow aquifers were tapped and bore wells with depths less than 30 m were very common. However, with increasing demand of groundwater and the influx of better and faster drilling technologies deeper aquifers were targeted and borewells of depth more than 100 m became common. It is only in 2019 that the Government of Maharashtra passed a law restricting the depth of borewells to 60 m, but its implementation is really very limited.

Photo 4.3 A close-up showing the dust that emerges due to the pulverization of basaltic rock while drilling through the dry hard compact basaltic flow unit

According to Duraiswami et al. (2012) due to the increase in popularity of borewells, the aquifers at deeper and deeper depths area being tapped and it is very common for a borewell to penetrate two to three lava flows. Such borewells may tap either the unconfined phreatic aquifers (compound and simple flows) or semi-confined aquifers (simple flows) depending on the nature and geometry of the lava flows.

In 2013, Duraiswami has tried to segregate the various Deccan basaltic aquifers, their geographic occurrence, their recharge capabilities, and the structure (dug well or borewell) for ground water exploitation they sustain.

The conceptual ground water system scenarios in the Deccan Traps by Duraiswami et al., though seem to fit quite well with the description of compound Pahoehoe flows and simple flows, the only problem being that the borewell sections as described by Duraisami are based on regional studies in that area, the description and terminology of the Deccan basaltic aquifers suggested is based on field observation of flows over large areas (Fig. 4.2).

This terminology is difficult to follow while correlating the bore logs based on the samples that are derived from DTH drilling rigs, collected on a fixed interval of 1.5 m, as in the case of the 120-mm-diameter borewells and 5.5 m in the case of 165-mm-diameter and above dimeter borewells drilled with the DTH rigs. Moreover, these samples as the drilling is in progress are usually collected by a person with little or no

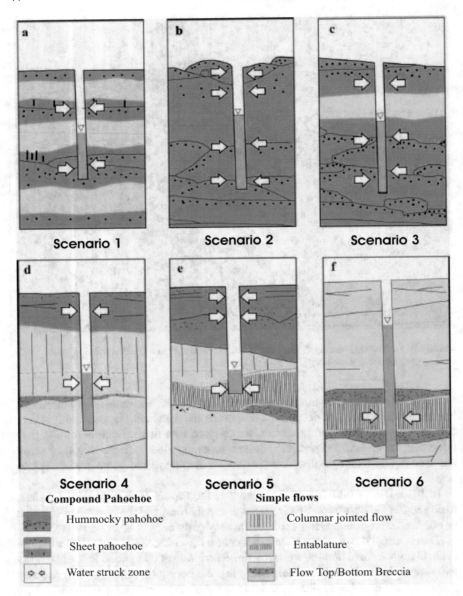

Fig. 4.2 Conceptual groundwater system scenarios in the Deccan Traps (**a–c**) compound Pahoehoe flows (**d–f**) simple flows (red stands—for vesicular basalt/flow top breccia and grey stands—for massive basalt). *Source* Duraiswami et al. (2012)

knowledge of geology. The problem regarding the use of this scheme of terminology suggested by Duraisami et al. (2012), being that the borewell sections as described by them are based on regional studies in that area, the description, and terminology of the Deccan basaltic aquifers suggested, is based on field observation of flows over large area, whereas the drill time samples that get collected while drilling are in progress and then taken to the laboratory for analysis and constructing a bore log, which makes it difficult for the person entrusted with this job to distinguish, whether the particular sample is from an Aa flow, simple flow, or a part of a compound Pahoehoe flow just by using these samples which are in the form of small chips or powder.

Even though the above conceptual models seem to be very attractive and descriptive, they are very theoretical and tend to make the whole regime look even more complex than it actually is, and it also makes it very difficult to understand thus making the task of correlation between two borewells within the same area a formidable task.

This task has become even more difficult in the present times where the new faster penetrating high-pressure hydraulic drilling rigs having a much higher number of impacts per second have replaced the older lower-pressure pneumatic drilling rigs due to which the material derived while drilling is of a much smaller dimension as compared to what it used to be while drilling with the older slower penetrating, pneumatic drilling rigs which used to operate at a much lower pressure, and hence, this kind of descriptive terminology even though sounds very nice and scientific from the genesis point of view and really does not prove to be useful from practical and field hydrogeological point of view.

The fact that drill time samples usually get collected at constant interval, which is dependent on the diameter of the well that is being drilled based on the length of the drill stem as the samples are usually collected every time there is a need to add an additional length of drill stem to drill further; there is a high probability that compact basalt flow units of smaller thicknesses which are a part of a larger compound flow or thin layered amygdaloidal basalt flow units sandwiched between two thick compact flow units tend to be missed out and not get recorded, and hence, the drill logs derived from such sampling do not really represent the true picture of the subsurface.

Many a times, the samples are collected using sieves which are placed in the water that is flowing out while drilling, and if the openings in these sieves are large, they fail to record the amygdaloidal basalt flow units and the red horizons as the samples derived from these are much finer in nature as compared to those derived for the compact basalt flow units, and once again, this would lead to a wrong interpretation regarding the thickness of the flows as well as the aquifers encountered.

Lalwani (1993), had observed that the terminology like "compound and simple types" or the "Pahoehoe and Aa types" can be used for the description of the exposures in areas such as road cuttings, hill slopes, stream cuttings, etc. provided a contiguous rock mass is available for observation. The compound flow units are composed of small units of compact as well as amygdaloidal types of basalts. It is rather difficult to distinguish between simple/compound types of flows and Pahoehoe/Aa types of flows even within open dug-well sections. Similarly, it is even more difficult to

Photo 4.4 Photos of The Compact basaltic flow unit underlined by Amygdaloidal basalt flow unit which is capped by Red layer. At Katraj Ghat, Pune

differentiate between compound and simple types of flows in the samples derived from the borewells drilled using down-the-hole hammer Drilling rigs.

The red tuffaceous horizons in saturated dug-well sections or in the down-the-hole hammer drilling rig samples are very easily identified, but it is very difficult to ascertain the nature of the underlying flow unit, whether compound, Pahoehoe or Aa type as both may be capped by red brecciated layers (Photos 4.4, 4.5, and 4.6).

Since the flow genesis of the Deccan basaltic flows is not of prime importance while mapping the yielding zones within the Deccan trap terrain Lalwani (1993) was of the opinion, the terminologies "amygdaloidal" and "compact" types of basaltic flow units are rather simple to use for the description of basaltic flows (e.g., only the amygdaloidal flow units are capped by red brecciated layers). Not only can these be easily differentiated while observing surface outcrops in road cuttings or in dug-well sections, but also while observing DTH drilling rig samples (Photos 4.7, 4.8, and 4.9).

In order to be able to correlate the description of rock samples that emerge at the time of drilling and the lithology observed while mapping the surface geology or along road cuttings and excavations, etc., the terminologies "amygdaloidal" and "compact" basaltic types/flow units seem to be a better choice. These terms are easy in describing both the surface and subsurface geology, and for correlating the borehole logs that are arrived at after studying the samples collected during drilling operations using the DTH drilling rigs, which obviously are the most commonly technique used to drill boreholes in most hard rock terrains.

Moreover, what makes the use of this terminology more appropriate is that these can easily be differentiated by anyone making observations even while drilling of

Photo 4.5 Photos of The Compact basaltic flow unit underlined by Amygdaloidal basalt flow unit which is capped by Red layer. At Mohmedwadi, Pune

borewells is in progress, this is possible even if they do not happen to have any knowledge of geology nor an educational background in hydrogeology. This differentiation of the basaltic flows into just amygdaloidal and compact makes it easy even for the person operating the drilling rig to understand the difference in the flow units that are encountered as the drilling progresses, who by the way in majority of the borewells that are drilled in India till date do not have any formal education in geology or hydrogeology.

Rana and Vishwakarma (1990) proclaimed that as the boles and bole-laded volcanic breccia act as important aquicludes between simple flows, both shallow unconfined aquifers and semi-confined aquifers are common, and they further go to state that aquifers under confining conditions within the simple flows are of rare occurrence within these regions.

According to Singhal (1997), the basalts unlike other hard rocks are unique in their hydrogeological characteristics, and unlike the granites the basalts form multi-layered aquifer systems which being quite similar to sedimentary rocks. Groundwater within the basalts is not restricted to fracture zones and joints which cut across the individual flows, but are seen to be flowing along the sheet joints which are developed along flow unit contacts.

From the drill time observations, the depths at which water is encountered represent the upper and/or lower contact zones between the amygdaloidal basaltic flow unit and the Compact basaltic flow unit. These zones are termed as *"INTER UNIT*

Photo 4.6 Photos of The
Compact basaltic flow unit
underlined by Amygdaloidal
basalt flow unit which is
capped by Red layer, Dive
Ghat, Pune

Photo 4.7 DTH sample of
Aumygdaloidal basalt flow
unit showing presence of
secondary minerals and
amygdales. *Source* Lalwani
(1993)

Photo 4.8 DTH sample of
Red horizon at the flow
contact of Upper Compact
basalt flow unit and lower
Amygdaloidal basaltic flow
unit. *Source* Lalwani (1993)

Photo 4.9 DTH sample of
Compact basalt flow unit.
Source Lalwani (1993)

ZONE" or in short "IUZ" (Kulkarni 1987; Kulkarni and Deolankar 1995; Lalwani 1993).

These "*IUZ*" are characterized by a larger percentage of amygdales and also by the presence of subhorizontal interconnected sheet joints. Such a sheet jointed IUZ's in amygdaloidal basaltic flow units acts as inflow zones within wells, and similar system of sheet joints can be observed in the borewells too as seen in the photos below of an open dug well and an uncased borewell drilled near it (Photos 4.10 and 4.11).

The multi-layered aquifer system which is inherent to the Deccan basalts is both laterally and vertically very heterogeneous in nature (Deolankar 1980; Kulkarni et al. 2000), making it very difficult for exploration of the potential aquifers there by making it really difficult while designing the system to achieve the desired recharge to groundwater.

CGWB (2000), CPWD (2002), and Gupta (2006), all recommend the drilling of Borewells to tap the aquifers to achieve artificial recharge in hard rock areas. The lithological controls on the development of transmissive sheet joints within

Photo 4.10 Dug well at Shirgaon village, Taluka Mulshi, District Pune Exhibiting Sheet Joints. *Source* Lalwani (1993)

Photo 4.11 Sheet joints in an uncased borewell at Shirgaon Village, Taluka Mulshi, District Pune. *Source* Lalwani (1993)

the amygdaloidal basaltic flow units have been highlighted by Kulkarni (1987). Lalwani (1993), on the other hand highlights the difficulty in predicting such lithological controls with the help of resistivity data and the limited Hydrogeological data collected in field at different locations. Given below is a conceptual drawing of the Deccan basaltic aquifers system suggested by Kulkarni et al. 2000 (Fig. 4.3).

Lalwani K. et al. (2015), in their paper "*Groundwater Recharge in the Basaltic Terrain for Rainwater Harvesting is possible Everywhere But Not Anywhere*", have

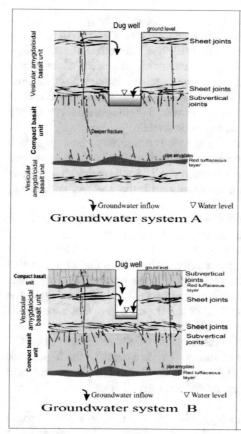

Groundwater system A consist of an upper vesicular amygdaloidal basalt unit and a lower compact basalt unit, in which the primary groundwater inflow zones are sheet joints in the upper and lower parts of the vesicular amygdaloidal basalt and subvertical joints in the uppermost part of the compact basalt.

Groundwater system B consist of an upper compact basalt unit and a lower vesicular amygdaloidal basalt unit, in which primary inflow zones are subvertical joints in the compact basalt and sheet joints in the upper part of the underlying vesicular amygdaloidal basalt.

Because of its well-developed network of sheet joints and subvertical joints, transmissivity and storage coefficient values are generally higher for aquifers in system A. Wells penetrating system A, therefore, are capable of irrigating more cropped land. The yields of large diameter open-dug wells can be classified based on this simple scheme. Groundwater potential of system A is better than that of system B, because wells tapping the former generally can irrigate a greater number of hectares in summer time per well.

Fig. 4.3 Basaltic aquifer system. *Source* Kulkarni et al. (2000)

highlighted the difficulties in achieving recharge to ground water due to this heterogeneity that is omnipresent within the Deccan basaltic aquifer systems. This, they say makes it difficult in identifying proper locations where the borewells would be in good hydraulic connectivity with the aquifer so as to be able to facilitate the recharge to the aquifer. They further go on to state that in hard rock terrain of the Deccan basalts, the borewells drilled for water harvesting for groundwater recharge need to encounter prolific aquifers and locations for such recharge borewells need to be properly identified along with the depth to which it must be drilled to be able to achieve the desired recharge too is important.

Multiple shallow recharge borewells (12–15 m in depth) drilled solely on the basis and positions of buildings and near the location of down-take pipes, as commonly practiced by engineers and architects and many rainwater harvesting Consultants, who are not adequately trained in hydrogeology while designing the rainwater harvesting layouts and who blindly tend to follow the standard design and suggestions made by CGWA, CGWB, or other such information which in current times are easily available on the internet, thereby keeping depth of the recharge well 2–3 m above post-monsoon water level which is actually meant to be for areas which have a thick weathered profile that is acting as an unconfined aquifer system within that area and not taking in to account the fact that it is not meant for areas where hard rock is exposed on surface and is also a part of the multi-layered aquifer system commonly observed to be occurring within the Deccan basaltic terrain.

Doing this in the basaltic terrain where hard rock is observed to be exposed on surface or occurring just below a shallow weathering profile of just a few centimeters in thickness will probably not achieve the desired recharge as the recharge borewells would probably not be having a good connectivity with the aquifers within that region.

It is commonly observed that the level of water in borewells and open dug wells usually rises a couple of meters above the inflow zones which are located within the horizontal sheet joints or the flow contact or the **IUZ** that are tapped by the well and which contributes to the hydraulic conductivity of the main aquifers within this terrain, and unless these are in hydraulic continuity with the borewells or wells, the structures in question will not be in hydraulic continuity with the aquifer within that area, as it is these IUZ comprising of sheet joints developed within that zone which are responsible for movement of water within the subsurface within the basaltic aquifers, which makes it essential that the recharge borewells penetrate these IUZ and go slightly below this depth and not be drilled only to a depth of the high water level in the area, which usually is not the static water level, but the piezometric level which is the level achieved due to the amount of formation pressure within the aquifer.

Even in situations where there is a thick layer of weathered zone above and if some recharge does take place, there is a high probability of this leading to a saturation of the near surface unconfined system, which would lead to the emergence of springs at its lower contact with the underlying hard rock and further recharge to this zone will only lead the increase in outflow from the spring at this contact and really not be of much benefit to the regional aquifer system in the region.

There is another major misconception prevailing that borewells which were dry and did not encounter any water while drilling are best suited for recharging. This probably is due to the fact that identifying locations for drilling borewells that yield water within the Deccan basaltic terrain is much more difficult than encountering a dry borewell. Hence, claims have been made that a "dry aquifer" has been encountered, which when subjected to recharge over a period of time, will start functioning as other normal aquifers and will start yielding water on a regular basis. An excellent example of such rainwater harvesting proving to be a total failure is that which has been executed at Prism Co –operative Housing Society Ltd, at Aundh, Pune (Photos 4.12, 4.13, and 4.14).

Going by the definition, an aquifer is a saturated geological formation capable of yielding sufficient amounts of water to wells. The basic function of an aquifer is storage and transmission; hence, a "dry aquifer" cannot exist. An overexploited

Photo 4.12 Borewell having very good hydraulic connectivity with the aquifer, Jeswani Associates site at Ravet, Pune

aquifer yields substantial quantities of water during the monsoon season, but very little during the summer months is very much a reality within the hard rock terrains. The fact being dry borewells did not encounter water while drilling are wells that are not in hydraulic continuity with the aquifer and hence not suitable for recharge purpose (Lalwani K. et al. 2015).

According to Kulkarni et al. (1997) and Lalwani (2000), borewells and dug wells in close proximity do not necessarily have the same hydraulic connectivity with the aquifer nor do they have the same yield; this fact confirms that Deccan basaltic flow units forming aquifers are extremely heterogeneous in nature due to the presence or absence of lateral interconnected joint system hence extreme caution has to be adhered while dealing with them. A completely dry borewell near high yielding borewells could be a case of very poorly developed or improper connections with the aquifer (Lalwani 1993).

This is something that really needs to be taken in to account when selecting suitable locations for drilling of recharge borewells, the computerized flow net mapping which is usually done to derive flow lines really do not give the true picture when used in

Photo 4.13   Borewell having limited hydraulic connectivity with aquifer at Khese Park site, Pune

the basaltic terrain; this is because of the heterogeneous nature (both vertical as well horizontal) and the multilayers nature of the of the Deccan basaltic aquifer.

While undertaking any scheme to try and induce recharge to groundwater system within the Deccan basaltic terrain, one needs to keep in mind this heterogonous nature of the Deccan basaltic aquifer and also the possibility of the borewells drilled not getting connected to the aquifers, or even when there is a very good connectivity the recharged water may not flow in the direction one expects it to flow and hence the intended recharge may not take place in a particular direction it was intended to while designing the system.

This makes it very essential to properly identify suitable locations wherein it is important for the recharging borewells to be in hydraulic continuity with the aquifer system as well as the benefit of the recharge being enjoyed by the entity for what the whole exercise of recharge was carried out, and for this reason the interconnections of the transmissive IUZ's need to be well developed and well understood to achieve the desired effect. Figure 4.4 is a generalized map of the Pune University Campus area; it clearly depicts the multi-layered nature of the Deccan basaltic aquifer, along with the lateral heterogeneity observed within the basaltic aquifer system.

It is because of this multi-layered heterogeneous aquifer system within the basalts that regional studies using satellite imageries and GIS application for assessment of potential recharge zone following the SLUGGER-DQL score, and other such tools and algorithms as attempted by Duraiswami et al. (2009) and Patil and Mohite (2014),

Photo 4.14 Dry borewell having no hydraulic connectivity with aquifer site at Kharadi, Pune

fail to provide a realistic picture on a local scale. This is due to the fact that they do not take into account the heterogeneity and the multi-layered nature of the basaltic aquifer into consideration, and many a times the conclusions regarding recharge potential areas within the basaltic terrain are misleading. Hence, one needs to be selective, where it comes to using such tools for evaluation of recharge potentials, more so when one working on a local scale.

Not only it is difficult to locate suitable locations where the aquifer could be well connected to the recharging borewell within the basaltic terrain, but one also needs to test the recharge capacity of the aquifer along with the storage capacity of the system to really understand the amount of water that can actually be recharged into the system at a particular location.

One of the best ways one can get some insight to the subsurface variations within the Deccan basaltic terrain without drilling is by using geophysics, viz. earth resistivity. According to Ratnakumari et al. (2012) in heterogeneous terrains too, earth resistivity imaging can successfully be used for identification of potential zones of groundwater. Lalwani (1993) is of the opinion that earth resistivity is the cheapest, best, and most applicable geophysical method for groundwater exploration. It was Conrad Schlumberger (1912) who first initiated the dynamic aspect of introducing

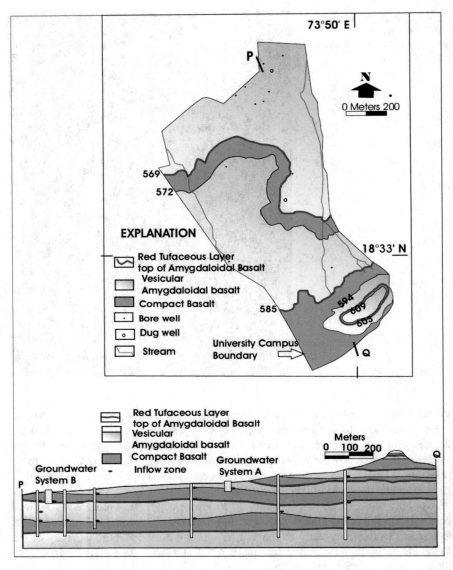

Fig. 4.4 Geological map of the University of Pune campus showing alternate layers of vesicular amygdaloidal basalt units and compact basalt units. Generalized geological section *PQ*, showing concept of multi-layered aquifer system in basalt terrain. Drill time well-log data were used to construct the section, although bore wells and dug wells shown in the section are at not exactly along the section line but a few meters on either side (Modified after Kulkarni et al. 2000)

the electric current into the earth to measure the electrical properties of the subsurface rocks. The porosity and the chemistry of water within its pores determine the
ability of a rock unit to conduct electrical current, and as the porosity, quantity of
water content, and the hydraulic conductivity and salinity increases the resistivity of
the rock units decreases.

Many workers just to name a few, Deshpande and Sengupta (1956), Banerjee
(1969), Bindumadhav (1974), Bardhan (1974), Limaye (1978), Zambre and Thigale
(1980), Khan (1987), Ghali (1990), Golekar et al. (2014), Rai et al. (2015), Patil et al.
(2015), Babar and Muley (2018), Desai et al. (2020), Navarro et al. (2020), have used
electrical resistivity method for the exploration of the Deccan basaltic aquifers for
selection of suitable locations for the excavation of dug wells or drilling of borewells.

In India, most of the government departments which are involved with development of groundwater rely on the use electrical resistivity as the only geophysical
method for the selection of sites for borewells, dug wells, and groundwater structures,
as it is the cheapest and the fastest geophysical method for groundwater prospecting
(Kelkar 1990).

Lalwani (1993), is of the opinion that the magnitude of the drop or rise in the
resistivity depends upon the degree of resistivity or conductivity of the portion incorporated within the depth of the penetration of the current. Even though one is not
able to quantify the drop in resistivity with the actual degree of saturation of the flow
unit as this factor is influenced by the lithological characteristics of the flow unit and
its hydrologic properties, it still can be used qualitatively to predict the presence of
a conductive basaltic flow unit within the subsurface. The depth at which the drop in
apparent resistivity is noticed is of great importance. According to Limaye (1978),
the depth of penetration of current varies between 0.5 and 0.33 times the current
electrode separations.

Apart from this, there have been studies involving remotely sensed data using
satellite imagery for groundwater prospecting, their effectiveness is limited to identifying lineaments and also to a certain extent for the mapping of the Deccan basaltic
flow units on a regional scale (Lalwani 1993). Making use of on-site earth resistivity
survey results one can get a better idea of the possibility of an aquifer and its depth
within the area which may be tapped via a borewell to be used to recharge the same
as compared to interpreting satellite based data in the laboratory.

Use of available computer programs for interpretation or one-dimensional or
two-dimensional resistivity modelling too does not really help much in selection
of suitable recharge locations. Especially, when the person who is interpreting the
field data is not the same person who has collected the field data, this is because
there is an overlapping values of the resistivity for the various different basalt flow
units depending on their saturation content, presence or absence of jointing and the
intensity of jointing and weathering.

A convention that is usually followed is to calculate the volume of the recharge
structure which usually comprises of a large chamber (2.25 m × 2.25 m × 1.75 m
with a dry borewell of about 20 m), which is approximately 10 m^3, which for all
practical purpose is considered as the recharge capacity per day of the system and
the number of such recharge structures depends on the quantity of desired recharge

that needs to be projected to be eligible for maximum points of IGBC or other Green Building rating agencies.

With no recharge capacity or infiltration tests being conducted, there is obviously no clear picture whether recharge if any is actually taking place. In most cases once the chamber gets full in the premonsoon shower itself and for the remaining monsoon season, the water flows out via over flow system, as the dry borewell drilled, are rarely in hydraulic connectivity with the aquifer, there is no water that gets accepted as recharge into the aquifers. What gets collected in the recharge structure would slowly percolate into the subsurface from the base of the chamber depending on the strata it is constructed within, or may remain stagnant until it is pumped out or until onset of summer where it tends to leak out from the side walls and evaporate as was the case in SOS Village site, behind Golf club, Pune, India.

It is such assumptions of recharge capacity, obviously, lead to wrong estimates of the recharge taking place over a month or year, and this probably is one of the reasons that in spite of so much efforts being made by the government to promote rainwater harvesting, the groundwater levels continue to decline with every passing year.

On paper, so many structures have been provided, but in reality probably no recharge really seems to takes place, or the recharge quantity that is estimated is far below the actually quantities that really land up entering in the system.

To be able to really understanding the quantum of recharge that is taking place through a structure one needs to measure the recharge capacity of the aquifer tapped by the recharge borewell that is drilled. There are many ways one can actually understand the recharge capacity of the aquifers tapped by borewells within the Deccan trap. One way of doing that is to consider the drill time yield as the rate of recharge the aquifer will be able to accept. This too has its limitation which in more ways than one is related to the hydrogeological set up of the basalts.

1. If the aquifer that was encountered at a shallow level, viz. 10–12 m while drilling and is located within an area where there is hardly any soil or weathered profile, underlined by compact basalt flow unit, then chances are that this aquifer will not be accepting much recharge. A good example of this is seen in one of the recharge borewells at Shilpa Cooperative Housing Society limited, Kothrud, Pune, and Maharashtra, India, where two of the three recharge borewells encountered nearly 10 m of weathered profile. The third borewell which was drilled at a slightly higher elevation encountered hard rock within 1 m of drilling. The drill time yield of this particular borewell, even though was much higher than the other two borewells, but when it came to accepting recharge the higher yield encountered while drilling was of no benefit and did not really help where it comes to accepting the rooftop water, as a matter of fact this borewell tends to start overflowing within a minute of the rooftop water being diverted into it for recharge purpose, what is even more disturbing is that this borewell tends to go dry in the summer months, when the need for ground water is at its peak. The situation in the other borewells within the same society premises separated by a maximum of 100 m is very different and in contrast to this the other borewells continue to accept

Photo 4.15 Large cavity in basalt observed near Chandini chowk, Pune

the recharge nearly till the end of the monsoon season and also yield substantial quantiles of water even during the dry summer months of April and May.

2. If the borewell while drilling encounters a gap and the drilling rod drops suddenly while drilling and a large quantity of water, much more than what is normally is encountered in other borewells in that area is encountered while drilling. (Photo 4.15 and Fig. 4.5).

   Under such situations too, it will probably be a deterrent to the borewell being successful in recharging with a capacity close to the yield recorded while drilling. The reason for this would be that this borewell has encountered a pocket while drilling and the yield recorded is not due to the hydraulic connectivity with the aquifer, but an anomaly which has occurred due to the presence of an isolated large cavity which probably does not have a well formed hydraulic connectivity with the regional aquifer. Borewell at Hotel Blue Nile, Bund garden Road, Pune Camp, Pune Maharashtra India, & Kamdhenu Earthmovers site at Mahalunge District, Pune, Maharashtra, India are good examples of this.

3. If the borewell has encountered water at depth e.g. 45–50 m, and the piezometric level recorded is only 10 m or less, indicating a rise of nearly 35–40 m, like in the case of "AAYAN", a project by Gandhi-Bafna, at Kharadi, on the Pune—Ahmednagar Highway just a few kilometers before Whagoli village & Marvel Edge, at Vimannagar, Pune. The reason for this being the formation pressure is

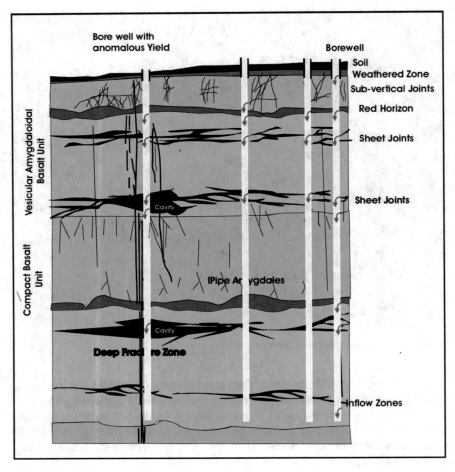

Fig. 4.5 Conceptual section showing borewell with anomalous yield

very high and does not allow water to be injected in to the system by gravity alone.

4. If the borewell is located on a ground which has a thick layer of black cotton soil, about 2–3 m and above, case of Jaganath Society, Boat club road, Pune and the water table is near about the soil to bedrock contact, then too it is not advisable to utilize this borewell for recharge purpose, as it is quite possible that the Black cotton soil will get saturated and will cause water logging around the borewell and could lead to formation "*kankar*".

5. At times the recharge borewells drilled in the summer months in an area where the groundwater system and the aquifers are already in a state of stress, wherein the aquifer is over exploited drilling in the summer months of April and May fails to yield any water or records very low yields of just 150 to 400 L per hour

Photo 4.16 Slug of water being injected for testing recharge capacity

while drilling, but surprisingly when subjected to recharging, these tend to accept a recharge at the rate which is quite substantial, and sometimes 10 times the drill time yield; hence, one really needs to thoroughly understand the state of the aquifer within a particular locality before deciding recharge capacity of any well that this drilled within the area.

Ideally, every recharge borewell should be tested for its recharge capacity; this can be done by measuring the radius of the well (r), standing water level (swl), and the elevation of the borewell Casing above ground (elv) and then pouring a slug of water in to the well at a rate equivalent to nearly 200 L per minute or more until the borewell starts over flowing, or an equivalent of 3000 L of water has been poured in to the borewell. Total volume of free space available in borewell xm^3 can be arrived at using the formula given below (Photos 4.16 and 4.17)

$$xm^3 = \pi r^2 \times [\text{elv}-\text{swl}]$$

Set the stop watch before injecting the slug of water. Once the borewell is full and starts to over flow, note the time required for filling the borewell at the specific flow rate, with this one can calculate the "y" the total volume of water that was injected in to the borewell (time $\times$ Flow rate = total volume injected = y m^3).

Using the formulas below, one can arrive at a fair estimate of the recharge capacity (RC) of the borewell

$$RC_1 = (y - x)/t \ \ m^3/\sec$$

Photo 4.17 Drop in water level being measured at fixed interval of time after slug injection

where

RC_1 = Recharge capacity of the Borewell in m³/sec.

y = Total Volume of water injected in m³

x = Total volume of free space borewell in m³

t = Time required for the borewell to overflow in seconds.

Another method to calculate the recharge capacity is that once the borewell overflows one should note the time required for it to start overflowing at the said injection rate. After which one should record the rate at which the water level drops until it reaches the level which is near about the original water level measured before pouring in the slug of water.

$$RC_2 = y/t_1 \ \text{m}^3/\text{sec}$$

where

RC_2 = Recharge capacity of the borewell in m³/sec.

y = Total volume of water injected in m³

t_1 = Time required for water level to drop to near about the original level in seconds.

Naturally, due to the dissipating water the rate of recharge measured is at a falling head and hence, the rate of recharge estimated will be lower than the actual recharge achieved at a constant head.

To get a better and a more accurate value for recharge capacity, one needs to carry out the slug injection while maintaining a constant the water level. The same procedure as described above needs to be followed; the only difference here would

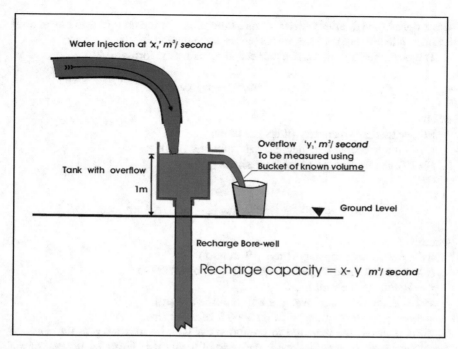

Fig. 4.6 Constant head recharge testing assembly for calculating recharge capacity

be that one needs to install a tank with an overflow spout system on the casing pipe as shown in Fig. 4.6.

When the tank gets full and the water starts flowing out from the sprout, the rate of injection is to be reduced to a point that there is no over flow from the tank but only from the spout provided for. This overflow can be measured using a bucket of a known volume and the time taken for it to get filled (average of 3 readings needs to be taken). It should be ensured that the water level in the tank is steady. Measure the injection rate (x_1), and also the constant rate of outflow from the spout (y_1).

The recharge capacity can be then calculated by using the formula

$$RC_3 = x_1 - y_1 \ m^3/\text{sec}$$

where

RC_3 = Recharge capacity (m³/sec).

x_1 = Injection rate of water (m³/sec).

y_1 = Rate of overflow from spout (m³/sec).

Occasionally, a slug of water comprising of 3000 L is insufficient to cause the borewell overflow; in this case one needs to note the time taken for injecting 3000 L (t_1), and note the level to which the water has risen in the borewell once the injection was stopped (swl_2), after which measure time taken for the drop in water level until it reaches near about the level it was before the injection (t_2).

The recharge capacity can be calculated by using the formula

$$RC_4 = 3000/(t_1 + t_2) \ \ m^3/\sec$$

where

RC_4 = Recharge capacity (liters per minute).

t_1 = Time required for injecting 3 m^3 of water (sec).

t_2 = Time required for water to dissipate (sec).

OR

$$RC_5 = 3000 - \left[\pi r^2 \times (swl_2 - swl)/t_1 \right] \ \ m^3/\sec$$

where

RC_5 = Recharge capacity (liters per minute).

t_1 = Time required for water injecting 3 m^3 of water (sec).

r = Radius of the well in m.

swl = Standing water level at the start of the test (m).

swl_2 = Water level after injecting 3000 L of water (m).

In most cases, the recharge to ground water is at its maximum at the onset of the monsoon as the aquifers are at the peak of their stress levels by the end of the summer months. This is due to the high volumes of withdrawal of water needed to supplement the Municipal supplies, and this apparently tends to be at its minimum during the summer months.

This rate of recharge slowly decreases until the time the aquifer has reached its peak saturation level. After which the aquifer usually does not accept any additional recharge, unless the system is still under stress and is being subjected to dewatering due to heavy pumping by wells in the neighborhood which are in hydraulic continuity with the system, and once saturation point is reached, the aquifers start to contribute to the base flow of rivers and streams. Once the base flow starts, it is an indication that the aquifer will no longer be able to store the additional recharge that is being injected into it artificially. This storage capacity of the aquafers within the region contributes to an important aspect while considering the sustainability of the aquifer as a source of water supply.

Apart from the difficulty of locating suitable sites for recharge, the fact that the Deccan basalts are alternate layers of amygdaloidal and compact basalt flow units with interflow units zones contributing to the secondary porosity and due to their higher transmissivity are acting as aquifers and the wells subjected to recharge at high elevation after reaching their saturation limits tend to flow out as Springs where ever the contact is exposed.

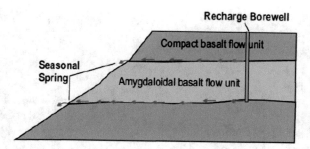

Fig. 4.7 Conceptual drawing of development of seasonal spring due to natural/artificial recharge (Modified after Source Pandit et al. 2018)

Photo 4.18 Spring at road cutting on Bombay–Bengaluru bypass near Pashan-Sus overpass, Pune

This development of springs at flow contacts which is inherent due to the multi-layered nature of the Deccan basaltic aquifers is observed all over within the Deccan basaltic terrain. At places these spring discharges have been channelized and been diverted to provide for the water supply schemes for some villages which do not have a good source by way of a perineal river or aquifer that they can tap within the low lying areas of their village boundaries (Fig. 4.7 and Photos 4.18, 4.19, 4.20, and 4.21).

Similarly, such springs perineal or seasonal are also found to be occurring at many places near about which most urban centers have been established. More than 35 springs have been identified to be occurring in and around Pune city (Kulkarni and Bhagwat 2019). The discharge from such springs that originate at lower elevations at flow contacts tends to be detrimental for aquifer storage, or one can say that once

Photo 4.19   Perineal Spring at Kule village, Taluka Mulshi, District Pune

Photo 4.20   Spring at contact of compact basalt overlain by Amygdaloidal basalt on Hill at Mugavde Village, Taluka Mulshi, District Pune

Photo 4.21 An ancient intricately carved spring seen behind Vitthal Temple on Sinhagad road, Pune India

the aquifer storage capacity is exceeded, the excess water is discharged as springs, which then contributes to the base flow of the rivers system. Such seasonal or perineal springs can be seen to be occurring at numerous locations all over the Deccan basaltic terrain.

All this points to the important hydrogeological characteristics of basaltic aquifer, viz. its storage capacity and the retention capacity and the question sustainability of the Deccan basaltic aquifers which needs to be studied, understood and taken in the account when coming to conclusions as to the ability of the Deccan basaltic aquifer systems contribution in overcoming the issue of shortage of water experienced in the summer months by artificial recharge.

References

Banerjee SL (1969) Geophysical investigations for groundwater near Belser, Purandhar taluka, Pune district, Geol Survey. India Report. Groundwater—Part I

Barber MD, Muley RB (2018) Investigation of sub-surface geology through integrated approach of geological and geophysical studies in the part of South-Eastern Maharashtra, India. American J Water Res 6(3):123–136

Bardhan M (1974) Application of nuclear logging and tracer techniques for location and evaluation of Trap aquifers in Maharashtra, India. Int. Symp. on Hydrology on volcanic rocks, Spain: 14.

Bindhumadhav V (1974) Geoelectric resistivity method as an effective tool for groundwater exploration in the Deccan Traps. Second World Cong I.E.R.A. New Delhi 3:246–257

CGWB (2000) Guide to artificial recharge for groundwater. Ministry of Water Resources, New Delhi, India

CPWD (2002) Rainwater harvesting—manual. Government of India, Nirman Bhavan, New Delhi110011. p 83

Deolankar SB (1980) Deccan basalts of Maharashtra—Their potential as aquifers. Ground Water 18(5):434–437

Deolankar SB, Mulay JG, Peshwa VV (1980) Correlation between photolinears and the movement of groundwater in the Lonavala area, Pune District, Maharashtra. J Indian Society Photo-Interpretation Remote Sens 8:49–52

Desai RV, Khan T, Gautam G, , Suryawanshi RA, Erram Vinit C (2020) Electrical resistivity investigation for groundwater potential in Lateritic Plateau of Bamnoli range, Satara District, Maharashtra. Bulletin of Pure Appl Sci 39F(2):264–273

Deshpande BG, Sengupta S (1956) Geology of groundwater in the Deccan Traps and the application of geophysical methods. Bull Geol Surv India, Ser B 8:27

Duraiswami RA (2013) Hydrogeology of aquifers in the Deccan Traps, India. J Geol Soc India 81

Duraiswami RA, Dumale VV, Shetty U (2009) Geospatial mapping of potential recharge zones in parts of Pune City. J Geo Society India 73:621–638

Duraiswami RA, Das S, Shaikh TN (2012) Hydrogeological framework of aquifers from the Deccan Traps, India: some insights. Memoirs Geol Soc India 80:1–15

Ghali SS (1990) A case study of Trappean and Kaladgi aquifers around Gadhinglaj. Unpublished Ph.D.Thesis, Univ. of Poona, District Kolhapur, Maharashtra, p 142

Golekar RB, Baride MV, Patil SN (2014) 1D resistivity sounding geophysical survey by using Schlumberger electrode configuration method for groundwater explorations in catchment area of Anjani and Jhiri river, Northern Maharashtra (India). J Spatial Hydrol 2(1):22–36

Gupta AK (2006) Rainwater harvesting. Indian Railways Institute of Civil Engineering, Pune, Maharashtra, India

Kale VS, Kulkarni HC (1992) IRS 1A and LANDSAT data in mapping Deccan Trap flows around Pune, India: implications on hydrogeological modeling. Archives Int Soc Photogram Remote Sen 29:429–435

Kelkar AD (1990) Identification of potential areas for artificial recharge by resistivity method and use of advanced geophysical technique for its implementation. In: Proc All India Seminar on "Modern Techniques of Rain water harvesting, water conservation and artificial recharge for Drinking water, Afforestation, Horticulture and Agriculture", Pune, pp 292–295

Khan IA (1987) Hydrogeology of Mahabeleshwar Plateau, Satara district, Maharashtra. Unpublished Ph.D. Thesis, University of Pune, 180 p

Kulkarni HC (1987) A study of the Deccan basaltic unconfined groundwater system from the Pabal area of Shirur taluka, Pune district, Maharashtra state. Unpublished PhD Thesis, University of Pune, India, p 285

Kulkarni H, Bhagwat M (2019) Pune's aquifers some early insights from a strategic hydrogeological appraisal, with Contributions from: Vivek Kale andUma Aslekar; Technical report ACWA/HYDRO/2019/H-86 ACWADAM, Pune

Kulkarni H, Deolankar SB (1995) Hydrogeological mapping in the Deccan basalts: an appraisal. J Geol Soc, India 46:345–352

Kulkarni H, Deolankar SB, Lalwani A (1997) Ground water as a source of urban water supply in India. In: Marinos K, Tsiambaos, Stournaras (eds) Engineering geology and the environment. Rotterdam, Balkema

Kulkarni H, Deolankar S, Lalwani A, Joseph B and Pawar S (2000) Hydrogeological framework of the Deccan basalt groundwater systems, west-central India, Hydrogeology Journal, VL - 8 : pp 368–378

Lalwani AB (1993) Practical aspects of exploration of Deccan basaltic aquifers for bore well development from parts of the Haveli taluka, Pune district, Maharashtra. Unpublished PhD Thesis, University of Pune, India, p 109

Lalwani A (2000) Study of deeper Basaltic aquifer(s) in Pune city & Environs, Maharashtra. CSIR–RA Report, No 9/137(249) M-EMR-I, council of scientific & industrial research, New Delhi

Lalwani A (2004) Efficacy of rooftop rainwater harvesting for artificial recharge to groundwater within the Deccan Basalts in Urban Areas (Abstract). Workshop on Rainwater Harvesting in Urban areas, Organized by Epicons Friends of Concrete, Mumbai

Lalwani K, Lalwani A, Mane B (2015) Groundwater recharge in The Basaltic Terrain for rainwater harvesting: everywhere but not anywhere. TERRE Magazine for Youth, Terre Policy Centre, Pune 1(1):2–6

Limaye SD (1978) Some aspects of integrated groundwater development. Unpublished Ph. D. Thesis, University of Pune, Pune, 260 p

Navarro J, Teramoto EH, Engelbrecht BZ, Kiang CH (2020) Assessing hydrofacies and hydraulic properties of basaltic aquifers derived from Kiang geophysical logging. Braz J Geol 50(4):e20200013

Pandit K, Lalwani K, Lalwani A (2018) Practical issues related to effective rainwater harvesting within The Deccan Basaltic province with special reference to the hard rock areas of Pune & Environs, Abstract Presented. G. D. Bendale Memorial National Conference on, Ground Water: Status, Challenges and Mitigation Moolji Jaitha College Campus, Jalgaon, India

Pathak MD, Gadkari AD, Ghate SD (1999) Groundwater development in Maharashtra state, India. In: 25th WEDC Conference on integrated development for water supply and sanitation, pp 192–195 Addis Ababa, Ethiopia

Patil SG, Mohite NM (2014) Identification of groundwater recharge potential zones for a watershed using remote sensing and GIS. Int J Geom Geosci 4(3)

Patil SN, Marathe NP, Kachate NR, Ingle ST, Golekar RB (2015) Investigation in Shirpur taluka of Dhule district, Maharashtra state, India. Int J Recent Trends Science Tech 15(3):567–575. http://www.statperson.com. Accessed 20 February 2022

Pawar SD (1995) Development and management of water resources in Dafalkhiwadi miniwatershed, Shirur taluka, Punedistrict, Maharashtra. Unpublished PhD Thesis, University of Pune, India, p 158

Rai SN, Thiagarajan Y, Kumari R (2011) Exploration for groundwater in the Basaltic Deccan Traps Terrain in Katol Taluk, Nagpur District, India. Current Sci 101(9):1198–1205

Rai SN, Thiagarajan S, Shankar GBK, Sateesh Kumar M, Venkatesam V, Mahesh G, Rangarajan R (2015) Groundwater prospecting in Deccan traps covered Tawarja basin using Electrical Resistivity Tomography. J Ind Geophys Union 19(3):256–269

Rana RS, Vishwakarma LC (1990) Occurrence of artesian conditions in the Sian River basin of drought prone Karjat taluka, Ahmednagar district in Maharashtra. Proc National Seminar on Modern Techniques of rainwater Harvesting, Water Conservation and Artificial recharge for drinking water, afforestation, horticulture and agriculture. Groundwater Surveys and Development Agency, Pune, pp 130–137

Ratnakumari Y, Rai SN, Thiagarajan S, Dewashish K (2012) 2D Electrical resistivity imaging for delineation of deeper aquifers in a part of the Chandrabhaga river basin, Nagpur District, Maharastra. India Current Sci 102(1):10

Singhal BBS (1997) Hydrogeological characteristics of Deccan trap formations of India, Hard Rock Hydrosystems. In: Proceedings of Rabat Symposium S2, IAHS Publ. no. 241

Singhal BBS (2009) Hydrogeological characteristics of Deccan trap formations of India. In: Proceedings, Hydrogeological co

Schlumberger C (1912) Premières expériences. Carte des courbes équipotentielles, tracées au courant continu Val-Richer (Calvados). Août-Septembre 1912. Ref 4717, Musée de Crèvecoeur en Auge, Calvados, France

Varade AM, Khare YD, Poonam P, Doad AP, Das S, Madhura K, Golekar RB (2017) Lineaments' the potential groundwater zones in hard rock area: a case study of Basaltic Terrain of WGKKC-2 Watershed from Kalmeswar Tehsil of Nagpur District, Central India. J Indian Soc Remote Sensing, December

Zambre MK, Thigale SS (1980) Geophysical investigations for groundwater in Sholapur District, Maharashtra. In: Proc 3rd Ind Geol Cong Pune, pp 435–447

Chapter 5
Long-Term Sustainability

Abstract Groundwater constitutes 48% of urban water supply in India. As per National Institute of Urban Affairs, 56% of metropolitan, class-I and class-II cities, are either fully or partially dependent on groundwater. The groundwater scenario in 28 Indian cities suggests that more than 50% of urban water is unaccounted for. Groundwater exploitation for commercial and domestic use in most cities is leading to reduction in ground water level. All this are indicators to the possibility of India becoming a water stressed country in the near future, with more and more people not having adequate access to clean drinking water.

Keywords Urban water supply · Groundwater exploitation · Water stressed · Water scarcity · Conservation · Rejuvenation

According to Narain (2012), groundwater constitutes 48% of urban water supply in India. As per NIUA (2005), 56% of metropolitan, class-I and class-II cities, are either fully or partially dependent on groundwater. The CGWB report of 2011 on the groundwater scenario in 28 Indian cities suggests that more than 50% of urban water is unaccounted for. Groundwater exploitation for commercial and domestic use in most cities is leading to reduction in groundwater level. All this are indicators to the possibility of India becoming a water stressed country in the near future, with more and more people not having adequate access to clean drinking water.

Amiraly et al. (2004), are of the opinion that water scarcity is a characteristic of north-western states of India, as a result of which over time people have developed techniques to meet their water requirements and one such technique is rainwater harvesting. In the old city of Ahmedabad, the rainwater harvesting system installed in ancient times was functional until the middle of the twentieth century that is before the extension of the modern water supply system to the entire town, this is true for all growing urban centers in India.

The continuous increase of the population has given rise to an increase in demand for water, due to which the focus has shifted on to groundwater to supplement the deficit. This in turn has resulted to an alarming depletion of aquifers. This led to the use of alternative sources of water, particularly of the rehabilitation of the rainwater harvesting structures still existing in its old city areas.

According to BASF (2020) in their online publication "India's urban water crisis calls for an integrated approach", an alternative rain water harvesting methods which involves harvesting rain water from concrete surfaces using porous concrete can be used to supplement rooftop rain water harvesting, to help replenish ground water.

Rain harvesting and conservation of water for augmenting the availability of water supply have been encouraged by The National Water Policy (2012), formulated by Ministry of Water Resources, River Development & Ganga Rejuvenation.

Similarly (Postel 1992; Reid and Schipper 2014) think it is necessary to explore the possibility of harnessing rain which is an easy accessible source to reverse the decline trend of water availability. Helmreich and Horn (2009) and Lange et al. (2012) consider rain water to have a very high potential for both domestic and agricultural use in the future as it is most easy to use and accessible sustainable water resource.

Rygaard et al. (2011) are of the opinion that rain water harvesting (RWH) systems are drawing global attention as an important component in attaining water security.

Abdulla and Al-Shareef (2009) and Srinivasan et al. (2010) suggest that there is unanimity where it comes to rainwater harvesting especially in urbanized locations experiencing severe scarcity.

In India too, it is during these last two decades, one has seen a surge in the revival of traditional practices of rainwater harvesting. This is being done to meet the ever-increasing shortfall of clean drinking water. Especially in the growing urban centers where there is a constraint of space, one cannot think of a common large reservoir, which in the past too, have really not proven to be very helpful, especially during the drought years and during the time the monsoon was delayed and rooftop rainwater harvesting seems to be the only way out for meeting the ever-increasing demands for clean water.

In the light of the above, it seems it is only right to really study the sustainability aspect of rainwater harvesting especially within the urban setup within the hard rocks of the Deccan basalts to try and understand the possibility of considering it as one of the means of draught proofing for water security in the future within the urban metros situated in the basaltic terrain.

To understand this, a few case studies have been selected; these particular cases were selected as they have been advocating rainwater harvesting as the wonder tool for eradicating the water scarcity which is faced by many in most of the fast-growing urban cities in India.

From what is being reported, they seem to have managed to overcome the issues of water shortage and are indicative of a probable development model for long-term water sustainability in the future.

Geologically speaking, these case studies are intricately related to the recharge and development of the hard rock aquifers of the Deccan trap area, and hence, the claims of success really need to be given a through scientific study, so as to understand whether they can be duplicated and implemented at other places too.

Many gated communities in the neo-development zones of urban India are faced with the problem of shortage of clean water, this is especially true in the summer months of April and May, some have this problem all throughout the year and need to depend of external sources of water supply, viz., water tankers, and this is because the local government does not have the capacity to provide for the fast-increasing demand both in terms of availability of resource as well as the infrastructure for distribution.

This lack of availability of clean water has led to the rampant and uncontrolled drilling of wells to tap the groundwater to meet the growing needs of fresh water within the urban centers which has led to the groundwater levels falling and borewells going dry as the aquifers getting overexploited.

The aquifers within Deccan basalts, as previously mentioned, are not easy to develop due to its heterogeneous nature as well as the low storage capacity and hydraulic connectivity, and hence, to try and stop this declining trend of water level, many have resorted to implementation of rainwater harvesting to try and recharge the groundwater aquifers any which way they think possible. The government agencies, namely the Groundwater Surveys and Development Agency (GSDA) of Maharashtra, which is a premier agency for groundwater within the state of Maharashtra, has been advocating rainwater harvesting for augmenting groundwater recharge for decades and believes that it is a sustainable solution for overcoming the problem of declining water table and water shortages in the region.

This crisis of water shortage in summer and the falling ground water tables has actually been a reason for many like-minded people coming together and registering themselves as a non-government organization to try and collectively tackle this problem and somehow find a practical solution to their common problem which they all face, which is lack of access to clean water. There have been some which have been formed with the sole intention of making money by selling some sort of concept, whereby their clientele are convinced that their problems regarding water will be get resolved if they do what these organizations or individuals advise them to, which in this case is being the implementation of rooftop rainwater harvesting.

As there are so many people involved in planning and implementing rainwater harvesting projects in India of which a vast majority really have no knowledge about hydrogeology and its importance, where long-term sustainability is concerned making it difficult to select case studies. It is of prime importance that a few select case studies, which seem to have achieved success and try to identify the reasons for their successes, have been chosen. The case studies that have selected are some of the most discussed and lauded rainwater harvesting projects which have made headlines in daily newspapers, their efforts have been recognized by the government and other organizations, and they have received various awards for the same and are being projected as role models for development by others facing similar problems.

Hence, trying to understand what exactly is the secret (if any) behind their success and how sustainable these systems really are on a long-term perspective and more important to understand whether it can be easily duplicated universally all over in the metros within the Deccan basaltic terrain where water shortage in summer is an ever-growing problem becomes a matter of prime importance.

5.1   Case Study I

5.1.1   The Greenland-2 Society a Gated Community in Vimannagar, Pune

The newspaper article in the daily newspaper "The Indian Express: Written by Nambiar Updated: April 5, 2016" speaks about how the members of this society especially Col SG Dalvi (retd.), an environmental activist heading the NGO "PARJANYA", and the chairman of the society were instrumental in implementing rooftop rainwater harvesting way back in 2003 by virtue of which the gated community managed to overcome the problem of water shortages they had been facing.

It is being claimed that before the implementation of the rooftop rainwater harvesting system in the society premises, the society was able to draw only a limited quantity of water from the aquifers that they had tapped via a borewell drilled in their society premises and the daily short fall of nearly thirty thousand (30,000) liters of water which they required was made up by exogenous source, viz., 3 water tankers a day, this meant an added cost to the residents.

The claim is that this shortage was overcome after implementing rainwater harvesting and diverting the rooftop water to recharge the aquifers, due to which the borewell started yielding additional water which was enough to meet their daily requirement.

This model is being promoted as a success story by Col Dalvi (retd.) to influence other gated communities who are facing water shortages during the summer month to undertake this exercise so as to be water sufficient. It is not only the common man who has been taken in, but due to these miraculous claims, the corporates and the government organizations too have started believing that rooftop harvesting is the solution to overcome water shortages faced in most metros during the summer months, which makes this as an important case study to be studied.

By simply splitting up the various components such as catchment, source, available water, one can easily arrive at some sort of logical explanation to the long-term sustainability of this system. Just comparing the amount of water that is being recharged to what is being extracted, one can visualize which way the water table will go.

Naturally, if the amount that is being recharged is more than what is being abstracted, it will have a positive effect on the groundwater levels, and that is what sustainability is all about, if not, then obviously this is not a sustainable solution.

1. The roof area of this society is about 1600 m^2
2. Rainwater harvesting potential (Yearly rainfall $\times$ catchment area $= 900\ m^3$ 700 mm $\times$ 1600 m^2 at 100% efficiency)
3. Daily groundwater abstraction 30 m^3
4. Yearly abstraction (30 $\times$ 365) 10,950 mm^3
5. Subtracting abstraction from recharge (10,950–900) the excess groundwater 10,050 m^3.

After looking at the recharge quantities and the abstraction that is being claimed, the questions that come to a mind of a scientifically oriented technical person in this case are:

1. Similar to seeds, does water tends to increasing in volume when diverted in to the ground?

If the answer is that it does not, which is very obvious even to a lay man, this then gives thought to the second question.

2. How does one account for the increase in the water that is being derived from the system?

One possible and logical explanation that comes to mind for this is that while trying to implement rainwater harvesting, they happened to tap the aquifer within their society at a place where the hydraulic connectivity of the system with the borewell was much better than what it was with their previous borewell, and hence, they were able to pump out the higher quantity of ground water.

The second possible explanation to this increase in quantity of available water is that there were some other issues with the pumping system, which seems to have been resolved during the implementation phase, and the borewell started functioning efficiently and could draw out the required quantity of groundwater from the same borewell.

From the above analysis of the data, it is obvious that the quantity of recharge is only about 10% of what is being withdrawn; hence, the success of being self-sufficient where water is concerned is not rooftop rainwater harvesting, but an obvious case of overexploitation of groundwater under the facade of rooftop rainwater harvesting.

It is also obvious that this system cannot be universally implemented due to the heterogeneity of the Deccan basaltic aquifer. It is defiantly not sustainable for the simple fact that more water is being extracted than is being recharged, this eventually would lead to the overexploitation of the aquifer, and hence, it definitely cannot be the long-term solution to the water crisis which is being faced by most housing societies during the summer months.

5.2 Case Study II

5.2.1 *Roseland Residency, Pimple Saudagar, Pimpri-Chinchwad, Maharashtra*

A newspaper article by Patwardhan, on May 9, 2009, paints a very rosy picture of the success of rainwater harvesting at the Roseland Residency, wherein the gated community was saving 100's of thousands of rupees after implementing rainwater harvesting within their premises and also was able to overcome their water shortage issues. Well, naturally because of the media exposure, the popularity of the society grew and it was conferred with awards for the work it has done and the success it has achieved.

The claim that the borewells, a total of 22 in the gated community premisis, were being used to supplement their daily water requirements, and during the summer months, some or all of them were not functional and they had to rely on external source of water supply through tankers, and after implementing rooftop rainwater harvesting, their borewells yielded substantial quantity of water and they saved the money that they had to spend on buying of water tankers.

As per the information that has been shared in the newspaper articles, we try to arrive at some conclusions regarding the various important aspects on a scientific basis and thereby can come to a fair conclusion as to the real success or failure of implementation of this systems on the basis of long-term sustainability.

1. Total area of the society 48,562.32 m^2
2. Number of buildings 30
3. Number of residents 2500
4. Daily water requirement 1100 m^3
5. Yearly water requirement 401,500 m^3
6. Daily availability from municipal supply 100 m^3
7. Yearly quantity of available municipal supply at 100% efficiency 36,500 m^3
8. Estimated rooftop area (as per 14,568.70 m^2 construction norms 30% of total area)
9. Yearly rooftop harvesting potential 9178.28 m^3
10. Total yearly water availability 45,678.28 m^3
11. Yearly excess withdrawal of groundwater 355,821.00 m^3.

From the above information, it is obvious that after using up all the water that can possibly be harvested, there is still a deficit which is being fulfilled by exploitation of groundwater. This essentially means that the groundwater system within that area is under stress and will slowly get depleted even when the area gets normal rainfall, and this obviously is not a sustainable solution, and similar to the first model, this model too is not one that can be easily duplicated everywhere due to the heterogeneous nature of the Deccan basaltic aquifer.

In this case too, it is very obvious that they are withdrawing a much higher quantity of water as compared to what they are able to harvest and recharge by way of rooftop rainwater harvesting. Here too, it seems that the quantity of water seems to have increased when injected into the ground for storage in the aquifers.

From years of experience of working in the field of designing, installing, and maintaining pumping system, what could have probably transpired that while drilling of borewells for implementation of rainwater harvesting, they have managed to drill at locations where the hydraulic connectivity of the borewell with the aquifer is much higher than what it was in the earlier drilled borewells, and they have managed to withdraw higher quantity of water from the existing groundwater aquifers which has helped them overcome the problems of water shortage that they had been facing. Secondly, there must have been some technical issue with the pump which, while executing the rainwater harvesting system, got resolved, the pump started functioning more efficiently, and they managed to withdraw a higher quantity of water for the same borewell.

Having said this, it is obvious that this system cannot be universally implemented on a sustainable basis, and it cannot be the solution to the water crisis being faced by most housing societies during the dry summer months.

The counter argument to this in both cases is that it does not matter as almost everyone is tapping groundwater to supplement their daily water supplies. This actually is true in most urban centers, but when there is the questions of whether it can be duplicated easily universally and whether it is sustainable and whether this system could be considered a wonder tool for overcoming the water shortages during summer months or in years where there is an extended summer due to the delayed monsoons and for preventing groundwater levels from declining, the above systems fail miserably on all counts.

This is being an unquestionable fact, and the reason for that being that the quantity of groundwater being extracted is much more than the quantity of water that is being claimed to be recharged, which apparently happens to be 100% of what falls on the rooftop; this, in itself, is an exaggeration, harvesting 100% of the rain that falls within a catchment is not possible due to the fact there are transmission, evaporation losses that need to be considered. Eventually, if everyone was to implement this kind of system, the aquifer would get overexploited and slowly dry out not due to inadequate recharge but due to overexploitation.

It is obvious here that not all information regarding the hydrogeological setup has been shared by the concerned persons, as they probably have no idea about it, and they have made claims that they have overcome their water scarcity issue simply by implementing rooftop harvesting within their premises. This in itself is very unscientific and hence cannot be considered as a universally applicable viable sustainable option to overcome water scarcity issues.

An article in the daily newspaper Times of India March, 2017, wherein it highlights the fact that of the Roseland success story which had influenced other gated communities in the neighborhood, viz., Kunal- Icon and Rosewood Apartments & Condominium co-operative society limited, who went ahead and implemented similar rainwater housing projects on their premises to recharge groundwater levels following the same methodology as by Roseland Residency Housing Society. This was done with a hope of getting similar benefit such as increased yield from their borewells, but unfortunately their efforts got them no positive results, and undertaking this costly exercise did not really get them any benefit in terms of additional water or saving on costs they were incurring for water, this just proves that what was implemented at Roseland Society is not universally applicable.

In another article, written by Marar, in the daily newspaper The Indian Express, Pune Published: July 3, 2019, which is about hiring 200 water tankers, and spending close to 2689 € during the peak summer months of April and May this year was a wakeup call for over 1000 families residing in Roseland Residency Housing Society in Pimple Saudagar, which a few years ago was claiming to have overcome their water shortage woes.

This, in spite of Pune getting higher than it normal rainfall which is 714 mm not only in the year 2018 but 2016 and 2017 as well (Deshpande 2019), is proof enough to support our above statement that the rainwater harvesting system implemented by Roseland Residency Housing Society is not really a sustainable one and would one day lead to the existing aquifer to be overexploited (Fig. 5.1).

To overcome this shortage, the society management felt the need for having better water conservation measures at the earliest. And hence, the society constructed its first recharge pit (9 feet × 12 feet × 13 feet), for harvesting surface water. This again seems to have helped the society in meeting its water requirements to a certain extent. How long this solution would last is a question that can be answered only in the times to come.

It is true that it is easy to dig pits and divert the surface water into the ground to recharge the aquifers, but something like this is not a sustainable solution to overcome water crisis in summer. It could help in meeting the high water demand for a few

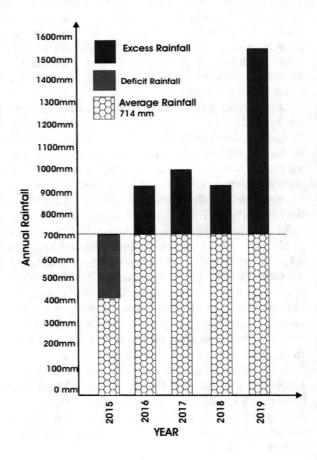

Fig. 5.1 Annual precipitation for Pune District (Modified after Deshpande 2019)

years, but in the long run, there is bound to be a shortfall in the quantity of water that can be extracted from an overexploited aquifer system.

At the same time, it is quite possible that due to this kind of recharge system being put in place, the surface contaminants would find an easy passage in to the aquifer system, leading to the deterioration of the ground water quality within the area, leading to other health-related issues to the users too.

A bare minimum water requirement of an individual is 135 L per day which is nearly 50 m^3 per year. To harvest 50 m^3 of water in an area where the yearly average rainfall is 0.7 m, an area of nearly 72 m^2 will be needed, this assuming an efficiency of 100%. In nature, it is impossible to achieve 100% efficiency, if it is reduced to 80% to account for evaporation losses, etc. Plus the vagaries in monsoon, the area required will be nearly 90 m^2. A similar figure has also been arrived at by Lalwani K. et al. (2015).

To implement rainwater harvesting to overcome water crisis on a long-term sustainable basis, the total catchment area that would be required to harvest water for 2500 residents to supplement their shortfall of 355,821.00 m^3 from the municipal supply would be equal to yearly water required divided by yearly rainfall, i.e., 355,821.00/0.7 this is nearly 512,428 m^2, which is more than 10 times the area of Roseland Society. This too, when 100% harvesting, is being considered, and this figure will be higher if we were to take in to consideration the losses one encounters in nature.

5.3 Case Study III

5.3.1 Banner-Pashan Link Road, Pune

Article by Khambete Posted Date: Sat, May 27, 2015. On the India water portal website.

A group of gated communities came together and collectively tried to find a solution for overcoming water shortage they faced in the summer months of April and May. The water was collected from the rooftops using pipes, was filtered, and then diverted to recharge borewells ensuring not a single drop of water was wasted. The flowing water from the hills was directed through a window fitted with a wire mesh on the walls bordering the hill to canals specially dug out to carry the water to the recharge pits for the borewells. Additionally, holes were made on the walls, so it does not come in the way of water from the hills reaching the canals.

The water collected on the rooftops and terraces was also channelized into the canals and also the storm water lines within the individual premises were diverted into the borewell recharge pits which were designed as per the GSDA as this whole exercise as mentioned in the article was undertaken under expert guidance hydro-geologists from GSDA, which is a premier institution established in 1972 by the

Maharashtra State Government, who conducted basic groundwater survey of the area and suggested locations for recharge.

ACWADAM, a leading scientifically oriented non-government organization, along with former deputy director (R&D) of GSDA, Dr. Shashank Deshpande, was involved in educating and convincing the society members regarding the recharge and discharge areas and also regarding the relevance of groundwater as a common property resource.

However just two years later in an article in the Times of India by Nair, Apr 3, 2019, Banner-Pashan Link Road in Pune hit by acute water shortage clearly shows that the rainwater harvesting that was attempted has really not benefited as much as expected. Members of gated communities on this road, viz., Orange County, Colina Vista, Mont Verte 1, etc. all have the same issues that borewells within the society premises have either dried up or are yielding insufficient quantity of water, and they need to rely on water tankers to supplement their shortfall in municipal supply of water.

For them, the whole exercise of implementing the whole system of rainwater harvesting was of almost no practical or monetary gain, as their situation remained the same as before with having to face water shortages due to inadequate supplies from their borewells.

The simple reason for this failure being, borewells were drilled in a channel through which all the water from hill slopes was diverted, with an assumption that as all the water from the hill slopes was being diverted through that channel and the recharge borewells drilled in it would automatically help in diverting this water in to the aquifers, thereby recharging the same.

This exercise proved fruitless as the borewells were randomly drilled within the channel, without trying to locate ideal locations in the whole area where the borewells could have been in better hydraulic continuity with the aquifers within that area. This was also done without trying to understand the capacity of the aquifers or the direction of the subsurface flow of the recharged water.

This resulted in the recharge either not taking place or the low-capacity recharged aquifers getting oversaturated and the recharged water flowing out at contacts at slightly lower elevations from where it was being recharged, without any benefit to the societies in that locality.

In this case even though some of the premier organizations and individuals who well versed with the hydrogeological setup within the trap and the problems associated with it were involved, they too failed to foresee the possibility of a failure by misjudging the direction of flow within the recharge borewells, which in the basalts poses a big challenge; this highlights the important point that was highlighted by Lalwani K. et al. (2015) that while working in the basaltic terrain, groundwater recharge is possible everywhere but not anywhere, which means that one should not attempt to recharge groundwater in the basalts with any presumptions or bias, but an attempt should be made to implement it in a place which is technically appropriate.

In this case though there was a scope of really doing a proper regional study, the recharge borewells were located only along the water channels, whereby they do not really encroach in to private territory of individual gated communities; this was done probably to give an appearance that they are being neutral and not favoring any specific gated community in particular, had this not been the sole criteria of selecting the recharge location they probably would have landed up doing some good and achieving better recharge to the system as a whole which could have benefit at least a few of the end users.

It is possible that instead of confining the recharge borewells to the channels along the base of the hill, they had been placed strategically at technically suitable locations, and the water from the channels diverted to these locations; the chances of achieving their goals of achieving groundwater recharge along with the societies getting the benefit of this would have been higher.

5.4 Case Study IV

5.4.1 Bhoomi Arkade, a 19-Story High Rise in Kandivali East Area of Mumbai

Article By: Karellia, August 23, 2019.

The society started with rainwater harvesting back in 2010. The rainwater collected on the terrace is sent to a tank on the roof through a pipe. The tank supplies water to the toilets of each flat. Throughout the monsoon season, even if there is an average rainfall of 10 mm, the overhead tank collects 13,000 L of water. Water is used for flushing and other non-potable needs of resident for 68 days, which is equal to the number of rainy days Mumbai has on an average in any rainy season.

In this case as the expectations were realistic, rainwater harvesting has worked successfully and it has helped them save on pumping and tanker costs for 68 days in a year when it rains.

Similar success stories where in the housing societies facing water shortages and are dependent on fulfilling this by being dependent of water tankers have tried to construct storage tanks and have utilize this harvested rainwater after some sort of filtration to supplement their water requirements. Naturally, this is not technically advised by either CGWB (2000), "Guide to Artificial Recharge for Groundwater" or CPWD (2012). Rainwater Harvesting and Conservation Manual, where in it is stated that surface storage tanks are not recommended where the gap between rainy and non-rainy days is more than 10–15 days as it involves considerable expenditure and the benefits of it are limited to the days, there is a downpour, in spite of this, the concerned end users are happy with what has been executed and they manage to save a lot of society funds which they would have to spend on procuring water tankers.

Rooftop rainwater harvesting can be a success if one does not link it to being a solution to the water crisis and reversing the effects of overexploitation of aquifers. It could be very useful and efficient in ensuring that the surface runoff generated during the monsoons seasons is minimized and water logging of large tracts of land avoided, it could also aid in reducing the hardness of water thereby improving the quality of water that is being pumped out of the borewells, especially during the rainy seasons, it could also be beneficial in reducing or completely stopping the ingress of saltwater along the coastal areas, but it definitely cannot be considered as a long-term universally applicable tool for overcoming the issue of water shortages during the summer months which is a common occurrence in metro cities with high-rise buildings not only within the Deccan trap area, but all over India.

Two other good examples of this can be cited one of which is a commercial building Continental Chamber, on Karve road, Pune, and the other is Ekveera Apartments, off law college road, Pune. An apartment building having just a 200 m^2 terrace, part of the ground was tiled and the other part is being used as a garden area, both these are multi-storied single buildings which were facing major water logging problems during the monsoon season within the building premises. The ground parking area was below the road level and with no avenue for the rainwater to drain out from the society premises. In case of Ekveera Apartments, during cloud bursts, the society being located at the lower most elevation of the lane, the runoff from the main approach road got accumulated within the society premises, and flooding in the parking area was a common of occurrence during the rainy season (Photo 5.1).

Coincidentally, both these sites have nearly 4–5 m of weathered profile and hence ideal for accepting recharge of the rooftop water. Nearly, 2–4 m^3 of water per day

Photo 5.1 a, b, c Rainwater harvesting at Ekveera Apartments off law college road, Pune; 2 out of 4 downtake pipes connected and diverted to recharge chamber adjoining the borewell

during the rainy days gets diverted into the ground through the recharge pit and borewell.

As a result of which, the parking is free of water logging on normal rainy days. The system tends to overflow whenever there is a cloud burst and there is a very heavy downpour, but tends to get cleared within minutes of the rains halting.

This can be termed as a success as it has achieved the target it was designed for, and it is also sustainable as the groundwater table is not being adversely affected. The borewells in this case are used only for gardening, cleaning purpose, or on occasions to supplement the water supply whenever there is a closure of municipal water supply, and there is no total dependency on the system to meet the high daily water requirement.

Hence, one can safely conclude that even though rooftop rainwater harvesting in urban centers is essential and beneficial in ways such as, it is helpful in reducing the runoff, thereby reducing the load on the municipal storm drains, it also helps in recharging the natural groundwater system which is it being deprived of due to urbanization, and last but not the least, it helps in improving the chemical quality of the groundwater by reduction in total dissolved solids especially during the monsoon and post-monsoon periods due to which there is lesser amount of scaling observed by people who depend on groundwater to supplement their daily water requirement.

On the other hand, rooftop rainwater harvesting even if it is achieved with 100% efficiency which, is next to impossible, cannot possibly be a sustainable solution for overcoming the water crisis in India's growing urban cities, wherein construction of tenements with such high density of residents that are permitted by the construction bylaws of the country.

It cannot be considered as a sustainable option to meet the growing water demand of the growing urban centers, especially not the ones within the Deccan basaltic terrain, where the multi-layered aquifer system is heterogeneous in nature both vertically as well as horizontally and of low storage capacity, and moreover, the rainy season is limited to a few months in a year, and the actual number of rainy days being less than 100 and the total hours that it actually rains is less than 150.

To achieve sustainability relying solely on isolated rooftop rainwater harvesting systems can never be a viable option, especially in regions where the rainy season is limited and the gap between the rainy and non-rainy days is large (10–15 days), and the total hours it actually rains is limited to around 100–150; it would require construction of huge reservoirs to store a few months of water supply, in an environment where there is already a shortage of open space to spread latterly. To add to this, there is the prevailing hydrogeological conditions which are non-conducive for uniform distribution of high amount of storage within the existing subsurface aquifer systems within the region.

If one expects not to be water stressed in the near future, there needs to be major change in the construction norms and proper preplanning needs to be done to evolve water smart cities in the future. There is a need to put restrictions on the spread of the urban sprawl and also in the permissible height of the buildings. To achieve success in evolving a water smart city, one needs to plan the limits of growth of a city on the basis of the available resources and not be based on a need to have extra housing to accommodate an infinite number of people. There is also an urgent need to convince people to conserve and reuse every drop of water, only then it is possible that Indian metros would be able to overcome their water woes and be less water stressed and more water sufficient in the coming years.

References

Abdulla FA, Al-Shareef AW (2009) Roof rainwater harvesting systems for household water 25 supply in Jordan. J Desalination 243:195–207

Amiraly A, Nathalie P, Singh JP (2004) Rainwater harvesting, alternative to the water supply in Indian urban areas: the case of Ahmedabad in Gujarat. In: IIMA Working papers WP2004-04-01. Institute of Management Ahmedabad, Research and Publication Department

Anjal M (2019) Jan Shakti Jal Shakti: Pune housing society digs recharge pit for water conservation. https://indianexpress.com/article/cities/pune/jan-shakti-jal-shakti-pune-housing-society-digs-recharge-pit-for-water-conservation-5812156. Accessed 10 October 2019

BASF (2020) https://www.basf.com/in/en/who-we-are/sustainability/future-perfect/stories/urban-water-crisis.html. Accessed 10 December 2020

CGWB (2000) Guide to artificial recharge for groundwater. Ministry of Water Resources, New Delhi, India

CGWB (2011) Groundwater scenario in major cities of India. Ministry of Water Resources, Govt. of India

CPWD (2012) Rainwater harvesting and conservation manual. Central Public Works Department, New Delhi

Deshpande S (2019) 1071.9mm: 2019 is Pune's third wettest monsoon in recorded history. https://www.hindustantimes.com/cities/1071-9mm-2019-is-pune-s-third-wettest-monsoon-in-recorded-history/story-yn6RFrXJAVzT63ZnYVWWXP.html. Accessed 10 October 2020

Helmreich B, Horn H (2009) Opportunities in rainwater harvesting, Presented at the Water and Sanitation in International Development and Disaster Relief (WSIDDR) International Workshop Edinburgh, Scotland, UK, pp 28–30. Desalination; 248 (1–3), Pub Elsevier B.V., pp 118–124

Karellia G (2019) Mumbai society harvests upto 13K litres rainwater/day. Saves Lakhs Using Solar Power. https://www.thebetterindia.com/192742/mumbai-society-sustainable-home-rainwater-harvesting-solar-power-saves-lakhs-india/. Accessed 20 December 2020

Kelkar K (2017) Their will, their way. https://www.indiawaterportal.org/articles/their-will-their-way. Accessed 20 December 2020

Lalwani K, Lalwani A, Mane B (2015) Groundwater recharge in the Basaltic Terrain for rainwater harvesting: everywhere but not anywhere, TERRE magazine for youth, TERRE Policy Centre. Pune 1(1):2–6

Lange J, Husary S, Gunkel A, Bastian D, Grodek T (2012) Potentials and limits of urban rainwater harvesting in the Middle East. Hydrol Earth Syst Sci 16(3):715–724

Nair A (2019) Baner-Pashan link road in Pune hit by acute water shortage. https://timesofindia.indiatimes.com/city/pune/baner-pashan-link-road-in-pune-hit-by-acute-water-shortage. Accessed 9 June 2021

Narain S (2012) Excreta matters, Vol 1. Centre for Science and Environment, New Delhi

National Water Policy (2012) Ministry of Water Resources, Government of India

Nisha N (2016) Pune: with rainwater harvesting, Greenland-2 Society says no to Tankers. https://indianexpress.com/article/cities/pune/pune-with-rainwater-harvesting-greenland-2-society-says-no-to-tankers/. Accessed 9 June 2021

NIUA (2005) Status of water supply, sanitation and solid waste management in urban areas. New Delhi

Patwardhan A (2009) A society in Pune is fighting water scarcity & saving ₹20 lakh every year. Here's How https://www.thebetterindia.com/99819/roseland-residency-pune-rainwater-harvesting-plantation-eco-friendly/. Accessed 9 June 2021

Postel S (1992) Last oasis: facing water scarcity. W.W. Norton and Company, New York

Reid H, Schipper L (2014) Up scaling community-based adaptation: an introduction to the edited volume. In: Schipper L, Ayers J, Reid H, Huq S, Rahman A (eds) Community based adaptation to climate change: scaling it up. Routledge, Abingdon, pp 3–21

Rygaard M, Binning PJ, Albrechtsen HJ (2011) Increasing urban water self-sufficiency: new era, new challenges. J Environ Manage 92(1):185–194

Srinivasan V, Gorelick SM, Goulder L (2010) Sustainable urban water supply in south India: desalination, efficiency improvement, or rainwater harvesting? Article Water Res Res. https://doi.org/10.1029/2009WR008698

Times of India (2017) https://timesofindia.indiatimes.com/city/pune/pimple-saudagar-society-uses-rainwater-harvesting-to-beat-water-woes/articleshow/57696782.cms. Accessed 10 October, 2020

Chapter 6
The Way Ahead

Abstract As mentioned in the previous chapter, if one expects not to be water stressed in the near future, there needs to be major change in the construction norms and proper preplanning needs to be done to evolve water smart cities of the future, this is especially true for the cities within the hard rock terrain of the Deccan basalts. Apart from the control of spread of urban sprawl both laterally and vertically, on really needs to evolve some sort of norm that needs to be adapted while constructing new structures.

Keywords Water stressed · Water smart · Hard rock · Construction norms · Urban sprawl · Shallow groundwater systems · Heavy dewatering

As mentioned in the previous chapter, if one expects not to be water stressed in the near future, there needs to be major change in the construction norms and proper pre-planning needs to be done to evolve water smart cities of the future, this is especially true for the cities within the hard rock terrain of the Deccan basalts. Apart from the control of spread of urban sprawl both laterally and vertically, one really needs to evolve some sort of norm that needs to be adapted while constructing new structures.

There is no doubt that groundwater will continue to be a supplementary or a complimentary and at times the only source of the urban water supply in India, even within the Class I & Class II cities. With so much having been done in terms of passing legislation, limiting the depth of drilling and curbing the abstraction of Groundwater by the infrastructure companies, and mining projects there still seems to be no real headway that seems to have been achieved by way of achieving water security, from what has been described in the earlier chapter, people have been indiscreetly withdrawing groundwater under the façade of reaping benefits of implementation of rainwater harvesting to recharge the groundwater system they are exploiting.

The local Municipal corporation development control rules, according to Nalawade et al. (2015), Draft Development Control Regulations for Development Plan Pune (2013) regulations 13.4.2(d) and 15.11, three Tier basements are permitted, which can be up to a maximum depth of 12 m. This legally permits the builders to excavate even further for laying the foundation of the building, which in turn at many sites lands up exposing the shallow groundwater system in existence within the trappean terrain. The existence of shallow groundwater systems (SGS) up to a depth of 20 m from ground level within the Deccan Volcanic Province is a well-documented fact (Lalwani 1993; Kulkarni & Deolankar 1995; Kulkarni et al. 1997). These shallow groundwater systems contribute to the primary source of groundwater that is usually tapped to exploit groundwater, and these excavations that are permitted by the construction by-laws according to Nalawde et al. (2015) have grave consequences.

The total amount of groundwater that gets wasted during the construction phase of an area of just 10,000 m^2. alone when the shallow groundwater system is exposed, is very high, Assuming a daily acceptable dewatering as per CGWA (2020) norms for infrastructure which does not require any detailed Hydrogeological investigation to ascertain groundwater impact, and post-construction the dewatering to ensure the basements are usable at such locations would be at least 365,000 m^3 per year, this is when assuming a conservative figure of only 100 m^3 per year average. In reality the figure is much more, especially during the rainy season.

After implementing rainwater harvesting in such an area with 10,000 m^2 of roof area in a city like Pune having 700 mm of rainfall will amount to just 7000 m^3 of annual recharge assuming 100% recharge is achieved, which leaves a deficit of nearly 358,000 m^3 and to replenish the groundwater that has been pumped out during the construction phase alone would require and equivalent of nearly 15+ years to be replenished. According to Nalawade et al. (2015), this figure is 32 years, as they have considered a more realistic daily average withdrawal of 500 m^3 which is based on data collected at various sites in and around Pune city.

Such dewatering of excavations has known to cause a drastic decrease in the amount of groundwater that can be pumped out from the borewells in the surrounding areas, examples of disruption of groundwater supplies have been observed at many location's in Pune city and suburb's, viz. Model colony, Yeshwantnagar, Subash-nagar, Lulanagar, Banner, Balewadi, Banner areas, where in the deep excavation for basement of buildings or underground metro station that are being subjected to regular dewatering and this in turn has drastically affected the shallow groundwater system in that locality. The effects of such as heavy dewatering drastically affect the performance of borewells in these areas; in some cases, the shallow borewells tapping only one aquifer going totally dry are of common occurrence; this is the case not only in the summer months but also during the monsoon seasons when the water table is usually at its highest level (Photos 6.1, 6.2, and 6.3).

Such heavy dewatering has a drastic and probably a non-revisable effect on the existing groundwater system within the hard rock aquifers in the region. The main reason for this is due to the fact that even though everyone knows that groundwater is being used to supplement/compliment the Municipal supplies in the urban metros all over India; groundwater has not been officially accepted as an important reliable, usable source of clean water within the urban centers by the government agencies.

While formulating the development control rules as in the case of Pune and Pimpri-Chinchwad Corporations, which usually is done by city engineers, an assumption that majority of the surfaces areas of Pune and Pimpri-Chinchwad area and other urban centers within the basaltic terrain having the Amygdaloidal basalt flow unit exposed on surface, which usually does not exhibit intense degree of jointing, and

Photo 6.1 Heavy inflow of water in excavation seen at Kohinoor B-Zone, Banner

that is the reason it does not have the potential to be termed as an aquifer, and only the compact basalts which exhibit intense jointing and cracks possess the secondary porosity and can form good aquifers and neither of these can sustain groundwater withdrawal over long periods of time.

Obviously, such grave and wrong interpretations would have been avoided if there was an authentic groundwater table map that was available, or a knowledgeable hydrogeologist was involved while formulating these rules.

It is not very uncommon that during the construction phase, no consideration is given to the depth of the water table within an area by the Architects while designing structure nor is it taken into consideration while granting permission for basements by the concerned authorities. The only mandatory clause is that the basement must be devoid of water to be usable, and that if dewatering is resorted to it should not be connected to the sewage system.

The only way it is possible to achieve this without the construction company having to resort to extremely high costs for water proofing is by dewatering, during and after construction and connecting it to the storm water system, which obviously is not the best or environment friendly way of doing dealing with it, and it also leads to wastage of large quantities of valuable groundwater resource.

Photo 6.2 Simultaneous
dewatering being done while
excavation is in progress at
site on Paud road, Pune

   The 2019 CGWA norms in India permit infrastructure projects to withdraw 100
m^3 of water per day for domestic use without any detailed environmental impact
assessment, and at the same time, it also permits up to 300 m^3 of water per day
for dewatering of excavation for construction purpose on similar basis. For higher
amounts though there is some sort of environmental impact assessment that needs
to be done, but such regional study in the hard rock terrains of the basalt are really
of no practical value due to the highly heterogeneous nature of the basaltic aquifer
(Fig. 6.4).
   The other issue here being that the rainwater harvesting schemes that get approved
by various environmental committees which again are comprising mainly of retired
or in service government personnel who by profession are mainly engineers and have
no qualified a hydrogeologist on board even when there is one like in the case of
GIRHA the person appointed to look in to the hydrogeological aspects is usually a
retired professor of geology, who's specialization is probably in some other branch
of geology and who has never worked in the field of hydrogeology but has acquired
his knowledge regarding hydrogeology by referring to some research paper or books
that are available, which really is not enough to understand the prevailing ground
realities. More so, when it comes to dealing with the enigmatic rocks such as the
Deccan basalts. Moreover, India is a large country with varied hydrogeology which

Photo 6.3 Deep excavation for multimodal transport hub, subjected to heavy dewatering during construction phase at Swargate, Pune

makes it impossible for any one single person to be able to understand the various hydrogeological regimes; this is because the groundwater dynamics are poles apart when one looks at soft rock hydrogeology and hard rock hydrogeology. Within the hard rocks too, basalts hydrologically are very different for the granites, schists, quartzites, etc., due to this lack of technical insight, the assumptions of recharge that are projected by the hydrogeologists hired by promoters get easily accepted, but in realty not even 10% of this assumed recharge actually takes place.

Naturally when there is no real testing of the recharge wells before they get approved such things are bound to happen. As Lalwani K. et al. (2015a) have mentioned in their paper "Groundwater Recharge in the Basaltic Terrain for Rainwater Harvesting: Everywhere But Not Anywhere" which makes it very important to locate the proper locations for drilling of borewells which would be in good hydraulic connectivity with the local aquifer system to be able to have a proper connectivity with the aquifers to be able to artificially recharge the Deccan basaltic aquifers.

There is an urgent need to limit the excessive dewatering of the deep excavations, which in my opinion is a far greater threat to groundwater depletion which is being observed within the first 10 m below ground, as compared to drilling of deep

Photo 6.4 Basement being subject to dewatering during the monsoon season at complex in Wakad area

borewells. For this there is a simple alternative; one needs to ensure that the depth of excavation is a minimum or may be just 50 cm above the existing aquifers within that area. If the aquifer does not get exposed, naturally there will be no need for dewatering during and after the construction and the groundwater that usually is pumped and wasted out on a daily basis from building having deep basements to ensure that they do not get water logged would be a thing of the past and naturally this would be one way of ensuring the groundwater is not depleted, and the groundwater table is maintained.

Secondly, the rainwater harvesting estimates should not be on the based on just surface and subsurface non-destructive testing, but it should be based on the actual results of borewells which have been drilled and tested for their recharge capacity with conceptual bore logs as shown in Fig. 6.1, which should made and kept on record for others to access.

Such a schematic bore-logs clearly illustrate the heterogeneous nature of the aquifer in that area and the difference in connectivity attained by the borewells that were drilled and the execution of rainwater harvesting structures design and volume planned accordingly. Restricting depth of drilling to 60 m in the state of Maharashtra too is also something that needs to be given a second thought, the reason for this is that there are certain areas where it is essential to achieve drilling beyond 60 m to be

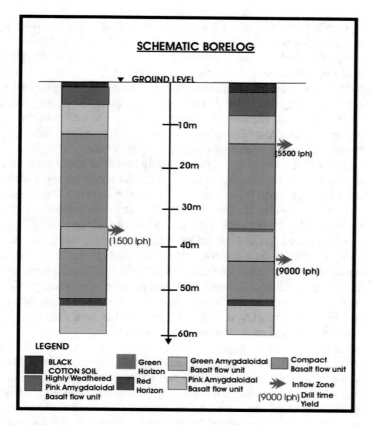

Fig. 6.1 Schematic bore-log of test borewells at Rama Builders Site, Bhoirwadi, Pune

able to be able to fully penetrate the aquifer and to derive benefits from if both for recharge as well as abstraction point of view.

IGBC (2019) along with the Ministry of New and Renewable Energy, Government of India, and The Energy and Resources Institute (2010)' GRIHA norms and other Green Building ratings norms too need to be given a second thought, and there should be no penalty by way of no points being awarded for no rainwater harvesting being executed at a particular site keeping in mind the practical issues related to recharge in the Deccan basalts, as it has been rightly pointed out by Pandit et al. (2018).

If by adhering to technical norms, no rainwater harvesting can be executed on that particular site it should be taken into consideration while calculating the green building rating, and the rating should be based on the percentage of points on the applicable factors, rather than the total number for all factors and zero points being considered for not implementing any rainwater harvesting, due to technical reasons. There should be no insistence of a storage tank to be constructed to store one-day rainfall, especially in regions where the gap between rainy days and non-rainy days is large, as it has been stated in the norms put forth by CGWB and NBC.

If weightage is given for following the set norms rather than the need to gain points promoters of housing societies and apartment buildings, etc. would not implement structures which really do not benefit the end user nor are really helping recharge the groundwater in the long run nor are they really helping replenishing the groundwater levels. This in turn would not lead to the end user having to incur huge operational and maintenance costs in the coming years, which is a big disappointment for them to keep the defunct rainwater harvesting system operational.

Moreover, points that are awarded for achieving sustainable rainwater harvesting should be based on actual testing results and not on some assumed recharge or construction of storage tanks to accommodate a quantity equivalent to one-day rainfall at nearly peak intensity for that regional unit which in this case is district. It is common knowledge that the precipitation distribution all over the district is not uniform; some parts of the district get very heavy rainfall, while others may be getting scanty rainfall as they fall within the rain shadow region or due to other micro-climatic variations; hence, designing a system in such areas where the rainfall is much less than the district average the assumption of the quantity of recharge taking place and the benefit being derived from it really are very much exaggerated as compared to the reality. Hence, it is important that the local rain gauge stations data is made available and easy to access when needed.

The aim should not be to achieve 100% recharge, but to achieve the recharge that was originally taking during the predevelopment days, may be a little more would also be acceptable, considering the fact that groundwater is been tapped to make up for what is not got from the Municipal supplies. Naturally, while trying to achieve 100% recharge, one may land up affecting the ecological balance and it could lead to disastrous consequences in the long run, especially in the downstream areas of the watershed. Ideally speaking, if recharge systems are designed based on the average daily precipitation of a particular area, one could safely assume that adequate recharge would take place and the recharge that is being done will not land up getting wasted by flowing out at flow contacts due to the oversaturation of the aquifers.

It is obvious that the government agencies and the green rating institutions do not have the adequate expert man power to assess every site, and under such circumstances, there will always be claims of much higher recharge taking than what actually happens all this just to gain some extra points, which should be made a thing of the past and there should be no extra points given for higher than required recharge, instead higher points should be granted for the quantity of excess treated water from the STP that normally gets disposed of through the sewage system, is treated to drinking water quality and reused not only for gardening, but also for domestic purpose within the project site.

In a country like India, where the norms are formulated by studying and comparing the norms and best practices followed in developed countries, without trying to first understand the ground realities that exist in India and how they are different from these developed countries, e.g., how the hydrogeological systems may be different from those in other countries and the absence of high density of rain-gauge stations, etc. More so, adherence to norms and executing the same as per the technically specified operating procedures being a standard practice in developed countries in contrast to that in India, the implementation is just to fulfil government building completion requirements without really bothering to achieve any real benefit to the environment.

Many a times, the norms set are for the whole country as such and cannot be implements at the local due to environmental diversity that exists, which has not been taken in to account while formulating the same or as there is a huge data gap regarding the prevailing ground situations, updating which in itself is a humongous task especially in a terrain which is dominated by hard rocks comprising of the Deccan basalts.

There is a need to really reassess the potential of groundwater as a source of water supply, and it should be given a complementary status rather than it being neglected and disregarded as it has been done in these past years. There needs to be a proper infrastructure put in place to ensure that groundwater being extracted due to any reason is done under strict regulations without any wastage and does not land up in the storm water systems, thereby adding to the existing load it was designed for, which usually leads to its failure resulting in water logging during the times it needs to operate on peak load capacity. It is essential that the groundwater be incorporated in a parallel piped water supply system within the area and distributed to others who could do with some extra fresh water.

Hopefully, in the near future the cities in India are not just Smart Cities with Wi-Fi connectivity and solar charging panels dotting the roads, but "Water Smart Cities", where it is easy to obtain basic information regarding the water resources availability in different areas, and more important being that the city is drought proof, wherein there is no indiscriminate wastage of the precious groundwater resources and the groundwater system helps in overcoming water shortages during water-deficient period which may occur as a result of deficient or delayed rainfall.

References

CGWA (2020) The gazette of India: extraordinary [Part II Sec:3(ii)]. Ministry of Jal Shakati, Notification, New Delhi, 4th September

IGBC (2019) Green homes rating system—version 3.0 for multi-dwelling residential units. Abridged Reference Guide September, IGBC, Hyderabad, India

Kulkarni H, Deolankar SB (1995) Hydrogeological mapping in the Deccan basalts: an appraisal. J Geol Soc, India 46:345–352

Kulkarni H, Deolankar SB, Lalwani A (1997) Ground water as a source of urban water supply in India. In: Marinos K, Tsiambaos, Stournaras (eds) Engineering geology and the environment. Rotterdam, Balkema

Lalwani AB (1993) Practical aspects of exploration of Deccan basaltic aquifers for bore well development from parts of the Haveli taluka, Pune district, Maharashtra. Unpublished PhD Thesis, University of Pune, India, 109 pp

Lalwani K, Lalwani A, Mane B (2015a) Groundwater recharge in the Basaltic Terrain for rainwater harvesting: everywhere but not anywhere. TERRE Policy Centre, Pune, India. TERRE J 1(1)

Ministry of New and Renewable Energy, Government of India, and The Energy and Resources Institute, (2010) GRIHA manual, Vol 1. TERI Press The Energy and Resources Institute, Darbari Seth Block, IHC Complex, Lodhi Road, New Delhi, 110 003

Nalawade P, Lalwani A, Lalwani K (2015) Lacunae in the formation of the urban DC rules detrimental to groundwater & its recharge Pune city, TERRE Policy Centre, Pune, India. TERRE J 1(1)

Pandit K, Lalwani K, Lalwani A (2018) Practical issues related to effective rainwater harvesting within The Deccan Basaltic Province with special reference to the hard rock areas of Pune & Environs, Abstract Presented. G. D. Bendale Memorial National Conference on, Ground Water : Status, Challenges and Mitigation "sponsored by North Maharashtra University, Jalgaon, Maharashtra; Orginised By: KCE Society's JalaSRI - Watershed Surveillance and Research Institute, Moolji Jaitha College Campus, JALGAON

Index

Printed in the United States
by Baker & Taylor Publisher Services